Natural Environments of Arizona

Natural Environments of Arizona

From Deserts to Mountains

Edited by Peter F. Ffolliott and Owen K. Davis

The University of Arizona Press   Tucson

The University of Arizona Press
© 2008 The Arizona Board of Regents
All rights reserved

www.uapress.arizona.edu

Library of Congress Cataloging-in-Publication Data

Natural environments of Arizona : from deserts to mountains /
edited by Peter F. Ffolliott and Owen K. Davis.
p. cm.
Includes bibliographical references and index.
ISBN 978-0-8165-2696-3 (hardcover : alk. paper)—
ISBN 978-0-8165-2697-0 (pbk. : alk. paper)
1. Natural resources—Arizona. 2. Plants—Arizona.
3. Animals—Arizona. I. Ffolliott, Peter F. II. Davis, Owen K.
HC107.A6N38 2008
333.709791—dc22   2008018870

Manufactured in the United States of America on acid-free,
archival-quality paper containing a minimum of 30% post-
consumer waste and processed chlorine free.

13 12 11 10 09 08   6 5 4 3 2 1

The royalties from this book have been donated to the
Arizona-Nevada Academy of Science for its scholarship programs.

Contents

Preface

Our understanding of the state's natural environments and eco-systems has increased greatly in the more than 40 years since Charles H. Lowe published his well-known book entitled *Arizona's Natural Environment* in 1964. In these intervening years, our knowledge of basic ecological processes and properties has expanded, and there has been intensified monitoring of the impacts of natural distur-bances, such as excessive rainfall events and devastating wildfire, and anthropological activities, such as increased urbanization, on eco-logical functioning. In addition, more easily accessed data sets and ecological information have become available to researchers, man-agers, and the general public. Our appreciation of Arizona's natural environments has broadened on all scales as a consequence. The aims of this book are to acquaint first-time visitors and the growing numbers of new residents arriving in the state with Arizona's spec-tacular natural environments, as well as to reacquaint long-time inhabitants of the state with these environments.

The authors have diverse professional backgrounds and experi-ences related to various aspects of the state's natural environments. The editors have attempted to retain the authors' interpretations of these environments in their respective chapters. The editors also treated each of the book's chapters as a stand-alone contribution, while retaining the common theme of Arizona's natural environ-ments throughout the book. Each chapter addresses a specific aspect of these environments, although there is some overlap among the chapters. The book is not necessarily intended to be a guide or handbook but rather to provide a basis for further study of Arizona's natural environments for those interested in doing so. Nevertheless, we hope that the book will be especially useful to people seeking a broader understanding and appreciation of the uniqueness of Arizona's natural environments.

Following an overview of earlier studies of Arizona's natural environments, the editors present a profile of the state in the book's introduction. D. Robert Altschul, in his chapter on the arid and semiarid lands of the world, places the state's natural environments into a worldwide context and perspective. William D. Sellers, Jon E. Spencer, Leonard F. DeBano, Daniel G. Neary, and Peter F. Ffolliott authored a sequence of chapters on the climate, geologic and landscape evolution, and soil and water resources that collectively shape the character of the state's plant communities. Descriptions of these plant communities and associations are the focus of a chapter by Peter F. Ffolliott and Gerald J. Gottfried. Diversities, compositions, and relationships of the vascular flora are detailed by Steven P. McLaughlin in his comprehensive chapter that includes analyses of the author's personal data sets. Habitats of commonly encountered animal species and efforts in sustaining the diversity of the fauna inhabiting Arizona's natural environments are reviewed by Paul R. Krausman and Peter F. Ffolliott. The editors conclude the book with a chapter on human impacts, which briefly describes how the state's natural environments have changed in the nearly 50 years since the publication of Lowe's milestone book.

Acknowledgments

The editors are grateful to numerous people in addition to the chapter authors for their assistance in completing this book. Many of these people provided valuable comments on early drafts of the chapters. We wish to acknowledge the staff of the University of Arizona Press, particularly acquiring editor Allyson Carter, for guiding us through the formative stages of preparing the manuscript, and senior production editor Alan M. Schroder for his assistance in the final editing of the book. We also thank our copyeditor, Lisa DiDonato Brousseau. Appreciation is extended to Cody L. Stropki of the School of Natural Resources at the University of Arizona, who formatted and prepared the figures in the book for publication. Finally, the editors thank the members of the Arizona-Nevada Academy of Science for their support in undertaking the preparation of this book to commemorate the academy's 50th anniversary in 2006.

Natural Environments of Arizona

1
Introduction

OWEN K. DAVIS and PETER F. FFOLLIOTT

The name *Arizona* is derived from the Tohono O'odham word *all shonak*, meaning "the place of little springs." The aridity of the state is proverbial and legendary. Some of the earliest prospectors were reported to have written home that, in Arizona, it was 80 feet to water and 2 feet to hell (Mann 1963). While the authenticity of this story is unknown, it attests to the one situation that conditions and circumscribes all of people's endeavors in the state—a continual shortage of water.

Many people think of Arizona largely in the context of its arid climate. However, the state is much more. Arizona is widely recognized for its uniquely diverse and picturesque landscapes. Much of the southern part is divided into isolated mountains and intervening desert valleys; a mountainous area lies diagonally through the middle of the state; and a high plateau is found in the northeastern corner. The most well known natural feature and a renowned tourist attraction is the Grand Canyon of the Colorado River, a fissure that is 215 miles long, 4 to 8 miles wide at its brim, and 4000 to 5000 feet deep. Ascending in elevation are mosaics of intermingling landscapes of deserts and grasslands, woodlands and forests, and, at the highest elevation, alpine vegetation. It was this large assemblage of diverse ecosystems that attracted C. Hart Merriam, an employee of the U.S. Department of Agriculture, to the Arizona Territory in 1889, where his well-known concept of life zones evolved and was later published in the "Results of a Biological Survey of the San Francisco Mountains Region and Desert of the Little Colorado in Arizona" by Merriam and Steineger (1890).

Early Studies of Arizona's Natural Environments

Merriam's concept of classifying plant communities by life zones was further developed and refined by Merriam himself between 1890 and 1900 (Merriam 1891, 1898) and continued by Mead (1930), Pearson (1933), McHenry (1934), Shreve (1936), Nichol (1937), Daubenmire (1943), Kuchler (1964), and others. Much of the collective work of these authorities dealt with verifying the utility of the life-zone concept in describing the natural environments of Arizona and—on a larger scale—elsewhere in the United States. In his widely read book *Arizona's Natural Environments*, Charles H. Lowe based his detailed descriptions of these environments largely on Merriam's life-zone system (Lowe 1964a). Lowe's book elaborated on the character of Merriam's life zones and described Arizona's biomes, biotic provinces, faunal areas, and floristic elements.

More recently, a hierarchical system of classifying plant communities was proposed by Lowe and David E. Brown. These researchers felt that a hierarchical system would be a more useful basis for classification because it would facilitate an orderly accumulation and entry of the continuously evolving information on the respective communities through time (Brown and Lowe 1980; Brown 1994). Structure of the biotic communities in this system is based on distinctive vegetation physiognomy within the communities. Although in general agreement with Merriam's life-zone concept, the hierarchical system of classification went far beyond Merriam's with mention of the original concept in only a largely historical context.

A Profile of Arizona

Arizona is the sixth largest of the United States, approximately 113,900 square miles in area. Of this total, 99.6 percent is land area with the rest surface water. A large portion of the state (almost 90 percent) is under some form of federal, state, or tribal jurisdiction. Privately owned lands are concentrated in central and southern agricultural regions and the two largest metropolitan areas of Phoenix and Tucson. Much of the former agricultural land has been—and continues to be—transformed into urban development as a result of the increasing influx of people into the state. Arizona's population has grown from about 1.3 million to nearly 5.8 million people since the mid-1960s, when Lowe published his book.

Climate

Arizona's natural environments are a product of their climate and geology. These environments largely follow the elevational gradients and temperature and precipitation regimes that characterize the state. From the lowest desert floors to the highest mountain tops, temperatures drop about 4 to 5°F for each 1000-foot rise in elevation, while annual precipitation increases from 3 to 4 inches per 1000 feet (Sellers 1960; Sellers and Hill 1974; Sellers et al. 1987). Overlaying these elevational trends in temperature and precipitation are atmospheric circulation patterns that provide the state with its unrivaled 300-plus days of cloud-free weather a year at the lower elevations. Cloud-free days are somewhat fewer at higher elevations.

Temperatures vary considerably throughout the state and, at given localities, the diurnal range can be extreme. Summer temperatures above 100°F in the shade are often registered in areas less than 3000 feet in elevation. However, these areas can be more comfortable than areas in the Atlantic and Gulf Coastal states because of the relatively low atmospheric humidity encountered. Summer temperatures above 7000 feet can be comparable to those that characterize southern Minnesota or Wisconsin. As an example of the diurnal range in temperatures that occurs in Arizona during the summer at elevations above 5000 feet, nights are cool and comfortable with temperatures of 50°F, while temperatures can exceed 95°F during the day. Near-freezing winter temperatures are occasionally recorded at the lower elevations, with low temperatures below 0°F for short periods following winter storms at the higher elevations.

Much of Arizona experiences two main precipitation seasons. A summer rainy season begins in late June or early July with the arrival of moist air from the Gulf of Mexico or the Gulf of California. Intense and localized thunderstorms often develop over the mountains and spread into the valleys during hot summer days, dropping up to several inches of rainfall locally. Winter precipitation arrives in the state from the middle of November through late March or early April, mostly in frontal storms from the Pacific Ocean. It commonly falls as gentle and widespread rains at lower elevations and snow on the higher mountains. Winter precipitation is usually the greatest during the periodic El Niño years. A greater portion of the annual precipitation falls in a summer monsoonal season as one moves southeastwardly, whereas winter precipitation is the norm as one moves southwestwardly.

Geology and Soils

The geologic features of Arizona provide an underpinning for the climatic patterns of the state. While the state's geologic history stretches back 1.8 billion years, its topography can be traced back to tectonic uplift, faulting, and erosion that began only 50 million years ago. A period of increased geomorphic maturity—including the integration of drainages into the Colorado, Gila, Bill Williams, and other large river systems in the state—has occurred in the past 5 million years. Increasing incision of the state's landscapes in the past 2 to 3 million years has been linked largely to climatic changes, with alternating ice ages and warmer periods. Aggregation continues to fill the basins inherited from earlier periods of faulting and volcanism in some areas. These processes have produced the topographic relief that is a major determinant of Arizona's natural environments.

Taken together, climate, geology, and topography shape the intricate patterns of soils throughout the state. Distributions of plant communities and associations are largely dependent on the characteristics of these soils and their inherent soil moisture regimes. Deserts are often regions of warm summer air temperatures and amounts of annual precipitation so limiting that soils are rarely moistened to depths greater than only a few inches. Grasslands, chaparral communities, and woodlands are found in areas of the state with prolonged dry periods and intervening but generally limited precipitation events. As a consequence, the soils fail to retain high levels of moisture for much of the year. Montane forests grow at higher elevations with comparatively greater amounts of annual precipitation, where the soils are typically wetter through much of the summer growing season and, to some extent, throughout most of the year.

Water Resources

Arizona's streams, which are largely ephemeral or intermittent in their flow, are fed mostly by widespread rainstorms of relatively high intensity that occur in the summer monsoon season or by snowmelt runoff in the spring. Only a few perennial waterways are found in the state, the largest of which is the Colorado River, forming much of the state's western border with California. The Colorado River has also given Arizona the Grand Canyon of the Colorado, one of the seven wonders of the modern world. Other rivers such as the Salt, Verde, and Gila originate in their highlands and, eventually, flow southward and westward into the Colorado River. The high-elevation streams that feed these rivers tend to flow rapidly over streambeds of coarse sediments, while the flow of water at lower elevations is more commonly sluggish and often muddy.

Increasing pumping of groundwater aquifers because of society's increasing demands for water has dried some streams and rivers of the state. Concurrently, the scarce water flows through watersheds often contain high concentrations of physical, chemical, and biological pollutants, impacting the quality of the remaining water supplies. With Arizona's continuously expanding population, high-quality water sufficient to meet future demands will remain a critical and often limiting resource into the future. Accelerated pumping of groundwater has compounded the problem of sustaining the supplies of high-quality water throughout the state.

Flora and Fauna

Arizona has an amazing variety of flora and fauna. The native flora consists of nearly 4500 species, with only California and Texas exceeding this number. The state's boundaries are not major barriers to the dispersal of plants, and no known floristic elements (groups of plants) are confined solely to the state. It is not surprising, therefore, that state-level endemism is relatively low. The number of naturalized exotic species is estimated to be about 410, representing about 10 percent of the total flora in the state (Brown 1994; Epple 1995). A relatively small proportion of widespread plant species exist in the state, with a much larger fraction of more narrowly distributed species.

Habitats for Arizona's diverse fauna range from sites with hot and arid conditions in the deserts and low-elevation grasslands to sites experiencing cool and moist conditions on the high peaks of the San Francisco Peaks near Flagstaff and White Mountains in the eastern part of the state. Overlapping habitats for large numbers of native and nonnative fish, amphibians, reptiles, birds, and mammals are found throughout these landscapes. Some of these species are rare, threatened, or endangered and, therefore, receive more managerial attention than the more commonly encountered species. Regardless of their official status, however, Arizona's faunistic resources are uniquely rich in their compositions, and sustaining their high level of biological diversity is a major aim of management.

It is within this general profile of Arizona that the climate, geologic and landscape evolution, soil and water resources, plant communities and associations, vascular flora, and fauna of the state's natural environments are described in the following chapters. Before these chapters, however, a general description of the arid and semiarid lands of the world is presented to place into perspective the discussions of Arizona's natural environments.

2
Arid and Semiarid Lands of the World

D. ROBERT ALTSCHUL

Arid and semiarid lands occupy nearly one-third of the Earth's land masses, an area roughly equivalent to North and South America combined and with a coastline longer than the African continent's (Middleton and Thomas 1997; Mainguet 1999). These lands include the hyperarid (extremely arid) lands that are often considered to be the "true deserts" of the world (fig. 2.1). Arid and semiarid lands share in common the broad characteristics of the climates, geologies and landforms, soils, hydrologic systems, and vegetation types that are associated mostly with arid environments and are distributed unequally in distinct geographic patterns among the continents (Heathcote 1983; Middleton and Thomas 1997). Arizona's natural environments are a part of the world's vast domain of arid and semiarid lands and their more humid borderlands, and perhaps no better example exists of the great diversity that characterizes the natural environments of the world's arid and semiarid lands than that encountered in the state.

Climates

While arid and semiarid lands exhibit a variety of geologies and landforms, soils, hydrologic systems, and vegetation types, they are defined primarily by their climates—or long-term atmospheric conditions—that result in broadly delineated zones of aridity when acting in concert with the Earth's surface. Arid and semiarid climates occur in two major and usually contiguous latitudinal zones: the tropics and the middle latitudes. Arid and semiarid climates in the

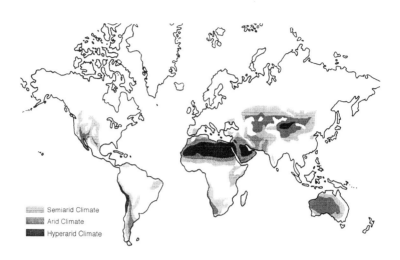

Figure 2.1. The world's drylands and zones of climatic aridity.

tropics are found in a zone from 15° to 35° latitude in both the Northern and Southern Hemispheres and extend generally from western coasts into continental interiors. This extensive zone of aridity experiences the high temperatures and high amounts of insolation with little seasonal changes that are associated with their tropical location. This arid zone borders lands with climates that have higher amounts of rainfall and more seasonally distinct rainfall distributions at their equatorward and poleward margins. On their equatorward margins, rainfall tends to be concentrated in the summer (high-sun) season, while rainfall on their poleward sides is concentrated in the winter (low-sun) season. Temperature contrasts between summer and winter in these arid zones are less equatorial and more distinct.

Middle-latitude arid and semiarid climates are found in a zone ranging from about 35° to 55° latitudes in both the Northern and Southern Hemispheres and tend to occupy the continental interiors. This zone experiences pronounced contrasts in temperatures and insolation between the summer and winter seasons. Precipitation follows seasonal patterns, and winter precipitation can occur as snow (Strahler and Strahler 1992). Similar to tropical arid and semi-arid environments, the bordering climates of the middle-latitude arid and semiarid lands influence the precipitation and temperature patterns in these regions. These patterns are complex, given the large latitudinal range and the greater prevalence of mountain ranges in these regions.

These precipitation and temperature patterns are well demonstrated in Arizona. The state straddles the generalized boundary between the tropical and middle-latitude arid and semiarid lands, and its climates and natural environments display elements of both. Arizona is also mountainous and has high plateaus in which elevation differences and slope aspects combine to produce diverse environments over short distances. Not all of Arizona is arid and semiarid in climate, however. More humid realms occur at higher elevations and often stand as isolated mountains surrounded by the more extensive drier lands. Both these upland and lowland realms share in the broad climatic systems that are characteristic of the state. A line drawn approximately north–south through the western third of Arizona marks the boundary between precipitation maximums in winter to the west and summer maximums in the east (Hecht and Reeves 1981). The transition from one precipitation regime to the other is marked by a transitional zone in which precipitation occurs in both seasons, as found in the Tucson area. This bimodal distribution of precipitation is a unique pattern in the world's arid and semiarid lands. It occurs in the state's lowland deserts as well as the more humid uplands. Similarly, great variability in annual rainfall with frequent droughts, wide diurnal temperature ranges, and other characteristics of arid and semiarid climates occur throughout the state.

Climatic aridity is caused throughout the world by interactions between the atmosphere and the Earth's surface below it (Goudie and Wilkinson 1977; Middleton and Thomas 1997; Mainguet 1999). The most prevalent cause of aridity is the presence of atmospheric stability and divergent circulation associated with high-pressure (anticyclonic) systems. These conditions are dominant in the tropical and subtropical latitudes centered in the vicinity of 30° North and South latitude and are known as the Subtropical Highs or Subtropical Anticyclones. These help to explain the presence of some of the world's largest arid and semiarid regions. Effects of these anticyclonic systems are also enhanced by continentality, where a continent's interior is separated from sources of moisture by great distances. Other causes of aridity are rainshadow effects and proximity to cool offshore ocean waters. Topography of sufficient elevation can act as a barrier to moisture-bearing air masses, causing orographic lifting and rainfall and snowfall concentrations on the windward side and arid conditions on the leeward side, where the preceding loss of moisture and the descending and heating air combine to minimize rainfall. The drying effect of a mountain barrier can extend for miles across the adjacent leeward-side valley floor.

Cool offshore ocean currents maintained by upwelling give rise to the coastal deserts of the world, which are some of the driest places on Earth. While sources of moisture are geographically close, the relatively cooler waters offshore produce strong atmospheric stability by cooling the lower layer of intruding air masses and, as a consequence, greatly inhibit the process of uplift and adiabatic cooling that is necessary for rainfall to occur. With some exceptions, cool-current deserts occur principally on the western coasts of continents roughly between latitudes 10° and 35° North and South. The extreme dryness of these landscapes is mitigated to some degree by the relatively high frequency of fogs, which provide moisture for flora and fauna.

The largest area of arid and semiarid climates is the Saharan-Arabian-Iranian-Thar Desert, a nearly continuous zone of tropical and subtropical arid climates extending from the Atlantic coast of North Africa 6000 miles eastward across the Middle East to northwestern India (fig. 2.1). Other major tropical and subtropical drylands occupy the bulk of the Australian continent, the Kalahari and Namib regions of southern and southwestern Africa, the Peruvian and Chilean west-coastal deserts, the southern interior and southeastern Patagonian region of South America, and the northwest-coastal and interior regions of Mexico. Arid and semiarid environments in the middle latitudes are located in two major regions: the interior of Asia, extending eastward from the Caspian Sea to Mongolia including the Turkestan, Gobi, and Taklamakan Deserts, and the interior of North America, principally within the western United States and northern Mexico (McGinnies et al. 1968; Meigs 1973; Goudie and Wilkinson 1977; Mainguet 1999).

Although climatic aridity is the common feature of these regions, major differences in the magnitude of aridity occur here, forming distinct geographic patterns. The three zones of climatic aridity shown in figure 2.1, hyperarid, arid, and semiarid, are recognized internationally (Middleton and Thomas 1997). Delineation of these zones is based on the groundbreaking works of Thornthwaite (1948) and Meigs (1953) in which the concepts of the soil-water balance and soil-water deficits are applied to a worldwide climate classification system. While the degree of aridity has been expressed as a function of precipitation and temperature in other climatic classifications (Baumer and Ben Salem 1985; Troyo-Dieguez et al. 1990), the zones of arid climates considered in this chapter are defined by ratios of precipitation to potential evapotranspiration. Water deficits are likely to occur when the amount of precipitation over an extended period of time in a region fails to meet the water needs of the atmosphere (evaporation), vegetation (transpiration), and water storage capacity of the soil. The greatest shortfall between precipitation and potential transpiration defines hyperarid zones. Hyperarid climates often form desert cores on maps depicting patterns of aridity, with arid and semiarid climates indicating a gradient and transition from this core toward more humid climates. Comparatively large variations exist in the distances over which the transition from dry to humid can occur in the arid and semiarid lands of the world.

The environmental impacts of aridity, the climates that account for them, and the human perceptions of them give unity to the descriptions and analyses of this large proportion of the Earth's land surface. Unique stresses and adaptations that are engendered by the locational, temporal, and economic scarcity of surface and subsurface water exist for plants and animals. Available water is the vital commodity in arid and semiarid lands, notable for its frequent absence or its localized and essentially nonrenewable and highly variable presence. Long-term water scarcity is a unifying and important component of arid and semiarid lands.

Although the arid and semiarid climates and associated levels of aridity provide an essential unity to the world's arid and semiarid lands, the diverse landscapes and environments found in these regions are not the exclusive result of today's prevailing conditions of aridity. Arid and semiarid regions contain a diversity of geologies and landforms, soils and hydrologic features, and plants and animals whose origins, ages, distribution patterns, and adaptability result in a variety of relationships with the arid and semiarid environments of today. Tectonic processes of the past and present, climatic changes that have occurred in past geologic periods and that are currently ongoing, the shorter-term effects of droughts and desertification, the natural preservation of relict landscapes, and adaptability to microhabitats help to explain the character of the world's arid and semiarid landscapes.

Geologies and Landforms

The land surfaces on which the world's arid and semiarid environments are found reflect endogenic and exogenic geologic processes that have operated in these regions over time and, as a consequence, produced a great variety of geologic structures, landforms, soils, and hydrologic features. Although the link between aridity and features such as perennially dry stream channels and sand dune formations seems obvious, no such connections are easily established with the vast array of land-surface and water-drainage forms that are found in most of today's arid and semiarid lands.

Desert Landscapes

Arid and semiarid lands have experienced different geologic histories and, as a result, contain a variety of geologic structures. These geologic histories are a product of endogenic forces whose energy derives from the Earth's interior and operate independently of the arid and semiarid climates found at the land surface. The geologies, land-forms, and other features of the two distinct deserts of the world—the shield-and-platform deserts and the basin-and-range deserts—are characteristic of the varying geologic histories and structures found in the arid and semiarid lands of the world (Hills et al. 1966; Mabbutt 1977; Heathcote 1983; Cooke et al. 1993).

The shield-and-platform deserts (also known as hamada-and-erg deserts and dome-and-basin deserts) occupy those parts of the continents that have been tectonically stable for long geologic periods. These desert environments have experienced denudation of younger rock formations, leading to the exposure of ancient continental rocks. They are partly exposed continental shields and partly vast sedimentary formations, far removed from the active boundaries of tectonic plates. Shield-and-platform deserts are primarily formed on the Gondwana Shields that comprise the continental and crystalline cores of Africa, western and central Australia, and eastern South America (Heathcote 1983; Strahler and Strahler 1992).

Topographic relief in shield-and-platform deserts is low to moderate, rarely exceeding a few hundred feet, with flat surfaces often extending to the horizon. These deserts are not without some mountainous relief, however. The central Sahara contains isolated volcanic areas such as the Hoggar and Tibesti Massifs, and the arid lands from northeastern to eastern Africa are structurally a rifted shield on which volcanoes and the broken terrain of the Eastern African Rift system add relief and diversity to the landscapes. Extensive areas of prominent topographic relief are rare in shield-and-platform deserts. Atmospheric processes are affected little by topography, and the seasonal movements of air masses, solar energy, and rainfall are largely controlled by latitude and distance. Topographic relief and elevation exert relatively little control over patterns of temperature, rainfall, and aridity. This type of desert structure dominates the Saharan-Arabian-Thar Desert, the interior Australian Desert, and arid and semiarid lands of southern and southwestern Africa. In combination with their regional landforms of dune fields, broad domes, vast and shallow basins, rocky platforms, and extensive stony surfaces coupled with the immense areas of these landforms, it is not surprising that shield-and-platform deserts are often considered prototypes of arid and semiarid landscapes.

The basin-and-range deserts (also known as mountain-basin-piedmont deserts and mountain-and-bolson deserts) are found in those parts of the continents that have been tectonically active in recent geologic periods, experiencing both the diastrophic and volcanic mountain building and structural shifting that are associated with the active zones of tectonic plates. These deserts are primarily associated with the tectonically active Alpine system of western North and South America; northwestern Africa; and western, central, and eastern Asia (Heathcote 1983; Strahler and Strahler 1992). Unlike the shield-and-platform deserts, comparatively great topographic relief, high elevations, and alternating mountains and structural basins dominate these landscapes. Atmospheric processes are not only controlled by the middle-latitude position of these deserts but are also influenced by the barrier effects and the elevations of mountain ranges. Rainshadow drylands result in some locations and middle-latitude forests in others—often in close proximity to each other and with a great variety of habitats between the two extremes.

Basin-and-range deserts typically consist of three interrelated landforms: the mountains that are frequently high enough to support more humid habitats, the piedmont zone that lies at the foot of the mountains and consists of alluvial fans and pediments, and the lower basins that are climatically the most arid parts but are also the beneficiaries of surface runoff from the more mesic adjacent higher zones. These topographic circumstances provide basin-and-range deserts with a combination of landforms, hydrologic conditions, and plant communities not normally present in shield-and-platform deserts (Peterson 1981; Cooke et al. 1993).

Basin-and-range landscapes dominate the arid and semiarid lands of North and South America, central Asia, and parts of southwestern Asia, primarily in Iran, Afghanistan, and Pakistan (Mabutt 1977; Cooke et al. 1993). Where mountains are less frequent and remote, as in the case of the Patagonian Desert, the Gobi, and the Colorado Plateau of northern Arizona, their structural landscapes can take on the appearance of shield-and-platform deserts and be so classified (cf. Mabbutt 1977 and Cooke et al. 1993). The classic display of basin-and-range deserts is found in the Great Basin of the United States, where large numbers of mountains and basins alternate in a pseudo-parallel pattern and, furthermore, where some mountains are high enough to support alpine environments more typical of high latitudes and display vertical zonation of plant habitats. The ratio of mountain-to-basin area varies greatly within this region, with basins generally occupying the greater amount of space. The higher mountains can be portrayed as islands separated by large areas of arid lands at lower elevations and are often called *sky islands* because of their relative floristic and faunistic isolation (Heald 1951 as cited by Lowe 1964a). It is the complex interaction of the subtropical and middle-latitude arid and semiarid climates with the basin-and-range structure that forms the basis for the great diversity of Arizona's natural environments.

Shield-and-platform deserts and basin-and-range deserts contain many distinct landforms and hydrologic features, most of which occur in both types of deserts but tend to occur more frequently in one or the other and vary between specific desert regions. Mabbutt (1977) identified six desert physiographic types in Australia, and Clements et al. (1957 quoted in Heathcote 1983 and Cooke et al. 1993) recognized 10 landform types and approximated their proportional occurrences in four major desert regions.

Other Landscapes and Surface Features

While endogenic processes provide arid and semiarid lands with distinct and varied structural outlines, the great variety of landforms or morphologically distinct surface features is largely attributable to exogenic processes whose energy, agents of denudation, and timing of occurrence are linked closely with Earth-surface climates. Weathering and the erosion and deposition of the weathered materials by fluvial and aeolian processes are dependent on available water. The accompanying arid and semiarid conditions have a significant impact on these land-forming processes. However, the link between the arid and semiarid climatic conditions of today and the origins of landforms that exist in today's deserts and semideserts is not obvious.

While many of these features tend to be described and analyzed under the heading of "arid and semiarid landforms," the links between these landforms and the processes responsible for them and the relationship of both to the prevailing arid and semiarid climates are complex (Cooke and Reeves 1976; Mabbutt 1977; Graf 1983; Smiley 1984; Cooke et al. 1993). While aeolian processes have produced landscapes that are most closely associated with the desert realm (such as the sand seas, or ergs), these are estimated to cover less than 30 percent of arid and semiarid lands of the world (Goudie and Wilkinson 1977; Cooke et al. 1993). Other aeolian features, such as rock yardangs, deflation hollows, and loess deposits in dryland margins, can be added to the aeolian domain.

Wind is an effective erosional and depositional force in many arid and semiarid regions, often abetted by people's misuse of soil and plant resources and contributing to processes of desertification. However, rarely are the landscapes devoid of the effects of flowing water on sloping landscapes and the accumulation of water in the lowest parts of basins. Stony and bedrock surfaces, such as regs, hammadas, serir, desert pavements, gibber plains, and gobis, that occupy large portions of the deserts of North Africa, Australia, North America, and central Asia are formed by the combined work of wind and running water. The finely chiseled surfaces of exposed rock formations and slopes, the fixed imprints of dry stream channels and drainage patterns that are evident in even the driest of locations, the accumulations of layered and sorted sedimentary deposits in the form of alluvial fans and bajadas, and the erosional surfaces of pediments all bear witness to fluvial activity. The presence of playas, salty lakes, and sebkhas, which are often bounded by abandoned shoreline features, further attest to the past, present, and intermittent presence of water in closed desert basins (Cooke et al. 1993). Most of the world's arid and semiarid landscapes show evidence of fluvial action, and past or present erosional and depositional fluvial processes continue to be the primary gradational force in the semiarid regions.

The widespread evidence of fluvial activity displayed by the land-forms of arid and semiarid lands seems to be at odds with the conditions of aridity that currently prevail in these regions. In large measure, the presence of these fluvial features in climates where fluvial processes are currently minimal, sporadic, and slow is evidence that past climates have been endowed with greater amounts of rainfall, snowfall, and the availability of surface and subsurface water. Areas can be landform museums in which the arid and semiarid conditions of today help to preserve relict features from more mesic times. Other evidence exists in geologic records, buried floristic and faunistic remains, and early human records of more mesic times in the distant past and within the last few thousand years.

Numerous studies have shown that arid and semiarid lands of today have experienced significant climatic fluctuations in the past 2 million years—roughly at the beginning of the Pleistocene Epoch including the past 5000 years (Martin 1963; Cooke and Reeves 1976; Mabbutt 1977; Graf 1983; Heathcote 1983; Smiley 1984; Beaumont 1993; Cooke et al. 1993; Goudie 1996). The nature of these changes, their frequency and timing, geographic occurrence, and effects are complex issues. It was during the Pleistocene that climates on what are today's arid and semiarid lands differed significantly, with evidence pointing to periods of more humid conditions than today in the hyperarid and arid cores and drier conditions in the semiarid margins. Evidence of permanent lakes with higher water levels marked by former shoreline features, the presence of paleosols and other deep weathered materials sometimes buried by subsequent deposition, and large river channels indicate that the Pleistocene experienced humid phases (pluvials). The presence of sand dunes currently stabilized by vegetative covers in semiarid and subhumid margins of deserts indicate that drier Pleistocene climates prevailed elsewhere. Post-Pleistocene climatic changes in time and space have been equally complex, with a general drying trend experienced by the central areas of today's arid and semiarid regions and wetter conditions in some of the marginal areas. Shorter climatic changes have occurred in the past 2000 years, so that the boundaries of the arid and semiarid lands continue to fluctuate, accompanied by changing geomorphic processes.

Soils

Soils of the arid and semiarid lands of the world are characteristically shallow, show little differentiation in depth (that is, horizonation), tend to have subsurface enrichments of minerals and salts that in more humid soil-water conditions would be dissolved, and are low in organic content in the more arid climates. It is common for "true soils" to be nonexistent over vast areas; if present, they are likely to be relics of earlier, more mesic times. The processes of physical and chemical weathering are handicapped in arid and semiarid environments by the paucity of water, and therefore soil formation in today's deserts and more arid climates is extremely slow. To the extent that soil accumulation through time is determined largely by the ratio of weathering rates to erosion rates, arid and semiarid lands would appear to benefit from the low frequency of runoff. But, except on relatively flat surfaces, the slow rate of soil accumulation is usually offset by its removal by occasional overland flows of water (surface runoff) even on gentle slopes. Exceptions to these characteristics exist in semiarid margins of the arid realm, where a history of greater and more reliable precipitation has produced soils with more horizonation and greater field capacities (Heathcote 1983; Strahler and Strahler 1992; Cooke et al. 1993).

Two soil orders, Aridisols and Entisols, are estimated to cover over 70 percent of all of the arid and semiarid lands in the world (Heathcote 1983 based on Dregne 1976). Both orders share an aridic soil-water regime in which soil water from precipitation is rarely available to plants. Aridisols have some horizon development and depth and have a high mineral and a low organic content; its suborders include soils with substantial clay or salt accumulations. Entisols display evidence of recent development, so that no horizonation is present due to either a lack of elapsed time or the repeated erosion of accumulated weathered material. One Entisol suborder includes soil-like accumulations that have formed on stabilized sand dunes. Entisols also occur as alluvial deposits on floodplains and deltaic areas in arid lands and can be highly productive.

Because of the comparatively lower variability and higher annual average rainfall, soil development has been more substantial in the more semiarid climates with deeper profiles, greater horizonation, greater accumulations of organic materials, and ustic soil-water regimes in which short rainy seasons can provide usable water for plants. Most notable of these soils are the Mollisols of the middle-latitude semiarid zone, which are among the most naturally fertile soils on Earth, and Alfisols in equatorial semiarid locations.

Soil types found in arid and semiarid regions reflect the local conditions of available water and quality of drainage, including sites where a combination of seasonal water accumulation, high water tables, and high evaporation rates are conducive to salt accumulations. Basin-and-range deserts contain a variety of soil types that is not matched by the shield-and-platform deserts in this respect. Although the soils of basin floors can share many characteristics with those in the shield-and-platform deserts, the transition from desert floors, through the bajada zone, to the mountain peaks can include soil orders whose physical characteristics and soil-water regimes have more in common with humid lowlands in other parts of the world.

Hydrology

Despite the conditions imposed by climates, the arid and semiarid lands of the world possess a variety of water sources and natural drainage systems that differ widely in their origin, geologic environments, dependence on climate, and spatial patterns (Goudie and Wilkinson 1977; Cooke et al. 1993). The origin of surface and subsurface water is either endogenic or allogenic. Endogenic drainage (surface runoff) is derived from precipitation occurring within the local area and therefore responds directly to the intensity, amount, and seasonality of local rainfall. The source of runoff in allogenic drainage systems lies outside arid and semiarid regions in either adjoining highlands or more distant humid regions. These drainage systems are more responsive to humid precipitation regimes, although they are increasingly influenced by local evaporation and infiltration rates in their progress across arid and semiarid landscapes. Many of the underground water reservoirs and groundwater aquifers of arid and semiarid lands are also considered allogenic in that their sources (intakes) lie in more humid areas. The slowness of water movement through its subterranean path shows up in the ever-widening gap between groundwater recharge and groundwater use in many human settlements. People living in the deserts and other arid and semiarid regions of the world frequently use ancient water, which for practical purposes is a nonrenewable resource (Goudie and Wilkinson 1977).

Drainage systems in arid and semiarid lands are also differentiated according to the terminus of runoff. Endoreic drainage terminates within these landscapes, often in large regional structural basins that lack surface outlets. Water losses due to evaporation and surface absorption contribute significantly to the ephemeral nature of streamflow and lake formation, giving rise to influent stream forms such as wadis, arroyos, and dry washes and collecting surfaces such as playas and salt flats. Sedimentation in desert landscapes is widely associated with the periodic flows in endoreic systems. In contrast, exoreic drainage normally reaches the ocean. The Nile, Niger, Colorado, Tigris, Euphrates, Indus, and Orange Rivers are streams whose flow in past times was maintained to ocean coasts despite heavy evaporative water losses en route. The total area of the arid and semiarid region that benefits directly from water in exoreic surface flow is negligible, however, and large areas display areic drainage systems, in which no discernable organization of drainage is evident (Mabbutt 1977).

Duration of streamflow in arid and semiarid lands is likely to be intermittent or ephemeral, in contrast to the perennial flows of even the smaller streams in humid realms. Intermittent or ephemeral streamflow regimes occur because of an almost exclusive reliance on precipitation events that determine the length and amount of channel flow. The tendency for channel beds to reside above the underlying water table so that channel flows lose water to the water table is another reason. Such streams are also known as influent streams. When streamflow events occur, their volume tends to diminish as the flow progresses downstream from the water source. Losses of water to evapotranspiration processes and infiltration of water into the channel beds can rapidly diminish streamflows as these flows proceed toward lower desert floors. Exotic rivers that manage to flow through desert landscapes on their way to the ocean experience great natural water losses. Flow duration and size of discharge in arid and semiarid land channels are determined directly by the frequency, magnitude, and timing of rainfall events and the position of the stream channel relative to the regional or local water table. The mesic environments of the higher mountains and the occurrence of snow cover during winter in the basin-and-range deserts of the middle latitudes provide streams with an opportunity of more prolonged flows, especially in the upper reaches of the drainage basin.

Vegetation

The plant life of arid and semiarid regions of the world defies simple description. Not only are there floristic differences among arid and semiarid regions, but the geologic, geomorphic, edaphic, and hydrologic circumstances of a landscape produce a multitude of habitats for plants with different growth requirements and adaptations. These plants include ephemeral annuals that appear after rains and complete their life cycle in a few weeks; succulent perennials, such as cacti, that store water for use in periods of drought through the enlargement of the parenchymal tissue and their low transpiration rates; and nonsucculent perennials. The majority of the plant forms found in many arid and semiarid lands are nonsucculent perennials. They include evergreen plants that remain biologically active throughout the year, drought-deciduous plants that become dormant in the dry season, and cold-deciduous plants that become dormant when temperatures drop below a certain threshold (McGinnies 1968; Goudie and Wilkinson 1977; Heathcote 1983; Strahler and Strahler 1992).

Many of these plants possess survival mechanisms that allow them to cope with the highly variable amounts and timing of water availability in arid and semiarid regions. Most plants of arid and semiarid regions are xerophytes, which tolerate prolonged periods of dry soil conditions and are opportunistic users of limited water supplies. The plants can also be grouped by their method of coping with variable moisture availability (Heathcote 1983), an approach that can be traced to the early work of Homer Shantz (Shantz 1927 as noted by Heathcote 1983). Drought escapers are usually ephemerals and annuals that emerge in response to a rainfall event and complete their life cycles within a short time span. Drought evaders are plants whose roots access a regular supply of water, usually groundwater. Often known as phreatophytes, drought evaders are often deep-rooted trees, shrubs, and palms and are common in riparian areas, where groundwater tends to be closer to the surface. Drought resisters are plants capable of storing water, such as cacti. Drought endurers are perennial plants capable of surviving in arid and semiarid climates by sporting small and waxy leaves, having large root systems in relation to the aboveground mass, reducing transpiration by closing their stomata during daylight hours, and absorbing carbon at night (C_4 plants). Some of the plants of arid and semiarid lands also have adaptations to saline soil-water conditions in high-evaporation sites, such as near salt flats and saline lakes. Classed as halophytes, these plants have also evolved mechanisms for coping with salt, such as evasion and toleration.

Nowhere are the profusion and variety of arid and semiarid plant life better displayed than in the natural environments of Arizona (Lowe 1964a; Hecht and Reeves 1981; Brown 1994; Epple 1995). Arizona contains parts of four deserts—the Sonoran, Chihuahuan, Mojave, and Great Basin—each with a distinctive combination of plant forms and species. Where the desert lowlands give way to ecosystems at higher elevations, other plant communities and associations appear, including grasslands, chaparral communities, and mixed woodlands of oak and pinyon-juniper. At the highest elevations, ponderosa pine, mixed conifer, spruce-fir forests, and, in rare instances, alpine communities are the predominant forms. As mentioned earlier, these communities are often portrayed as sky islands, separated from each other by the more xeric lowland communities. Riparian associations and wetlands occur throughout Arizona where water availability permits. At the lower elevations, these streamside communities can occur far into Arizona's deserts, where they are dependent on a variable and uncertain water supply.

Although elevation and its concomitant increase in available moisture is a dominant factor in explaining the great diversity and general distribution of these plant communities, the combination of distinct thermal and precipitation seasons—modified by mountain and basin topography with its varied orientations and elevations and a history of environmental changes in past millennia—produce a great variety of plant communities and associations and a complex pattern of boundaries between them. Increasing urbanization and second-home construction, ranching and other forms of land use, and the introduction of exotic plant species and other plant invasions have greatly added to the diversity and complexity of Arizona's vegetation and, therefore, the natural environments of the state. In this respect, Arizona is a microcosm of the global subtropical and middle-latitude patterns of vegetation and vegetation changes, which are represented in the state by pronounced elevational differences in an arid and semiarid land context.

3
Climate

WILLIAM D. SELLERS

An understanding of Arizona's natural environment is facilitated by an understanding of the climatic factors that ultimately impact and control its character and diversity. Precipitation and temperature regimes vary widely over time periods ranging from seasons to years, decades, and millennia, and these temporal effects are amplified by the spatial arrangements of local and regional topographic features such as slope, aspect, and differences in elevation. The book *Arizona Climate* (Sellers and Hill 1974) contains an informative presentation of the relationships of weather elements to pertinent topographic features of the state. Also included in this book are illustrations showing surface- and upper-air patterns associated with the heaviest summer and winter precipitation events. A more general discussion of Arizona's climate is presented in this chapter.

The Southwestern Climate Pattern

Much of Arizona is characterized by the Southwestern Climate Pattern, which consists of cycles of winter precipitation, spring drought, summer precipitation, and fall drought (Reed 1933, 1939; Jurwitz 1953; Smith 1956; Sellers and Hill 1974; Van Devender and Spaulding 1979). The two dry periods that occur in much of the state can limit the development of natural environments. The spring drought extending from mid-April through late June is usually more stressful to most plants than the fall drought lasting from mid-September to mid-November. Not only is precipitation generally less in the late spring but temperatures are also higher than in the fall. Flagstaff, located at the base of the San Francisco Peaks in northern Arizona, averages almost twice as much precipitation in September through November (5.27 inches) as it does in April through June (2.51 inches). The average daily temperature in Flagstaff is 4.4°F lower during the fall (43.8°F) than it is during the spring (48.2°F). Corresponding values of precipitation and temperature for Gila Bend, about 170 miles southwest of Phoenix, are 0.36 inches and 78.4°F, respectively, in April through June and 1.36 inches and 74.3°F in September through November. The fall drought is occasionally broken in September or early October by precipitation events that are associated with weakening hurricanes or tropical storms moving northward off the western coast of Mexico and Baja California.

Temperature and Precipitation Patterns

Daytime temperatures in the deserts of southwestern Arizona can exceed 100°F on more than 100 days of the year, accompanied by only about 12 days with measurable precipitation. At the opposite extreme, nighttime temperatures on the Mogollon Plateau in north-central Arizona can drop below freezing on 200 or more days a year, and measurable precipitation can occur on up to 100 days, with as much as 35 inches of precipitation for the year. While the daytime maximum temperature generally decreases with elevation from the low valleys to the high mountains at a relatively steady rate of nearly 4°F per 1000-foot increase in elevation, the early-morning minimum temperature is more variable and can increase with elevation over relatively short vertical distances—especially following clear, calm, and dry nights. This phenomenon is attributed to the tendency of the coldest and densest air to pool at the lowest elevations, resulting in strong temperature inversions.

Much of Arizona receives about as much precipitation in winter as it does in summer. The only exception is in the southeastern part of the state, which receives most of its annual precipitation in the summer. For example, the ratio of December through March (winter) precipitation to that of July through September (summer) averages 0.63 for Tucson and 0.46 for Bisbee. Even though winter precipitation is usually more widespread and fairly uniform in intensity over the state, it is more variable in amount and time of occurrence than summer precipitation (McDonald 1956). Arizona's statewide average winter precipitation has ranged from less than 1 inch to more than 12 inches during the past century (Sheppard et al. 2002). In the extremely wet winter of 1992–1993, the statewide average precipitation from December through March was 14 inches. The average precipitation for the same interval in the winter of 1998–1999 was only 1 inch. This variation occurs primarily because winter precipitation events depend largely on the location and movement of large-scale atmospheric circulation patterns that normally do not favor Arizona.

Precipitation Regimes

The geographic distribution and sustainability of Arizona's natural environments are strongly affected by both annual and seasonal precipitation regimes. Superimposed on these varying precipitation regimes are infrequent tropical storms that have a more pronounced influence on the hydrology of the landscape than on the longer-term stability of these natural environments.

Annual Precipitation

The total annual precipitation in Arizona is extremely variable. During the period from 1953 to 2002, total annual precipitation in Phoenix, situated in the desert of central Arizona, varied from a low of 2.82 inches in 1956 and 2002 to a high of 15.23 inches in 1978. During the same time interval, annual precipitation in Flagstaff varied from a low of 10.37 inches in 1956 to a high of 36.59 inches in 1965. The difference in total annual precipitation from the driest to the wettest year was much greater for Flagstaff (26.22 inches) than for Phoenix (12.41 inches). However, the ratio of the total annual precipitation in the wettest year to that in the driest year is greater for Phoenix (5.40) than for Flagstaff (3.53). In terms of ratios, therefore, variations in annual precipitation at the higher and wetter elevations (Flagstaff) are generally smaller than those at the lower and drier elevations (Phoenix), while the reverse is true in terms of total annual amounts. Which of these two measures of annual precipitation is a better indicator of the fragility of the state's natural environment depends largely on what is being studied. An unusually dry period impacts sites at low elevations by creating water shortages and lowering water tables and at high elevations by increasing the risk of wildfires.

Winter Precipitation

Winter precipitation is normally associated with airflow from the southwest in advance of a strong upper-level trough at 10,000 to 30,000 feet altitude moving eastward from the Pacific Ocean into the western United States. Often accompanying this trough is a surface low-pressure center that forms in the Gulf of Alaska, moves down the western coast of North America, and eventually swings inland over southern California. The largest and longest-lasting winter rainstorms occur when an upper-level low breaks off the main trough over the Pacific Ocean and stagnates off the southern California coast for several days, picking up heat and moisture before finally moving inland. However, this sequence of events is not common. More frequently, the system enters the continent in the Pacific Northwest, producing heavy rainfall in that part of the country and leaving the southwestern United States dry. Arizona is most likely to receive precipitation from these storms when the surface ocean-water off the Pacific Coast is warmer than usual, as often happens during El Niño events (see below). The warmer water and additional heat energy associated with these events have a three-fold impact on winter weather patterns. This combination allows the trough to extend further south along the Pacific Coast than usual; it contributes additional moisture to the atmosphere through increased evaporation; and it weakens the normally strong coastal marine inversion that traps most of the moisture and associated air pollutants below its base and prevents much of this moisture from moving across the coastal mountains and into the desert regions to the east.

El Niño

The importance of El Niño events and their opposite La Niña events on Arizona's climate cannot be overemphasized (Rasmusson and Wallace 1983; Philander 1990; Swetnam and Betancourt 1990). El Niños are initiated by a weakening of the persistent trade winds in the subtropics between $10°$ and $30°$ latitude in both the Northern and Southern Hemispheres. The cause(s) for this weakening are not clearly understood. The trade winds—driven by strong equatorial solar heating—can be weakened in some cases by a reduction in the solar radiation reaching the lower atmosphere due to debris injected into the upper atmosphere by a major volcanic eruption (Mann et al. 2005). The trade winds of the Northern Hemisphere blow from the northeast, while those of the Southern Hemisphere blow from the southeast. Because of the effect of the Earth's rotation, the trades drive surface layers of the underlying ocean away from the equator and westward from about the International Date Line eastward to the western coasts of North and South America and away from both western coasts from the equator to about $30°$ latitude. The displaced water is replaced by the upwelling of cool and nutrient-rich water from depth. The upwelling is especially intense during La Niña events, when the trade winds are stronger than normal.

When the trade winds weaken at the onset of an El Niño event, upwelling is suppressed and cool surface water is replaced by warmer and less-fertile water flowing eastward along the equator. At the same time, the region of maximum convective activity and low surface pressure normally over Indonesia is displaced eastward to a position at about the International Date Line. The result is that the atmospheric pressure falls and precipitation increases in the eastern equatorial Pacific Ocean and along the western coasts of subtropical North and South America. Concurrently, the pressure rises and drought occurs in the western equatorial Pacific Ocean and Indonesia, northern Australia, and India. The reverse occurs during La Niña events. The oscillation of high and low surface pressure between El Niño and La Niña events is strongest immediately south of the equator and thus is usually referred to as the Southern Oscillation. It is customary to speak of ENSO events in references to El Niño, emphasizing the close association of El Niño and the Southern Oscillation (Ropelewski and Halpert 1986).

Summer Rainfall

The summer rainfall season, often referred to as the summer monsoon season, usually begins dramatically in late June or early July, when the dry and warm air that typically prevails over Arizona throughout May and most of June is suddenly replaced by more humid and slightly cooler air (Xu et al. 2004). While this air mass commonly moves into Arizona from the Gulf of Mexico and the subtropical Atlantic Ocean, it can also enter the state from the Gulf of California and the subtropical Pacific Ocean, especially in late June and once again from late August through early October. The source of the monsoon moisture has been the subject of much debate (Douglas et al. 1993). It appears that the major source of this moisture varies from summer to summer, depending largely on the air circulation aloft. Regardless of its source, summer rainfall is mostly convective in nature and likely to occur in intense, short-duration events (fig. 3.1).

The most intense and widespread thunderstorm activity is usually associated with surges of very moist air up the Gulf of California (Hales 1972) when an upper-level high with counterclockwise rotation is situated over the four corners area of Arizona, New Mexico, Utah, and Colorado (Mitchell et al. 2002; Lorenz and Hartmann 2006). The warm and moist air from the south is destabilized as it passes over the heated desert surface. Intense thunderstorms, usually with diameters of less than 3 miles, occur when this unstable air ascends by convection or topographic lifting on the windward slopes of the mountain regions in the state. Daytime thunderstorms normally form in the early afternoon over the higher mountains and spread out over the surrounding valleys in the late afternoon and early evening hours. Summer rainfall is occasionally increased and covers a larger area than usual when a weak upper-level disturbance of tropical origin passes over the state from the east.

Figure 3.1. A convective buildup of clouds before the onset of a summer rainfall event.

The timing, spatial distribution, and intensity of summer rainfall events vary greatly from year to year. These precipitation events can occur in the early or late summer, or can be spread evenly throughout the season. In some years, there are periods lasting about a week with widespread thunderstorms occurring throughout the state. In other years, much of the rain will fall from one or two unusually heavy storms, with the rest of the summer relatively dry.

Tropical Storms

Tropical storms, while infrequent, represent extreme rainfall events in Arizona (Thorud and Ffolliott 1973). These storms usually form off the western coast of southern Mexico from late August through early October and move slowly northward and northwestward, occasionally intensifying to hurricane strength as they approach Baja California. They build mainly on the energy of warm ocean water. Therefore, when these storms pass over the cold California Ocean current flowing southward off the western coast of North America, they are normally weakened or deflected westward toward the Hawaiian Islands, depending on the direction of the upper-level flow. Even when this flow is from the southwest, the effect of these tropical storms on Arizona is usually minimal. However, when the current is weaker than usual and the surface waters are warmer than usual—as during El Niño episodes—some of the moisture can move into Arizona to produce widespread and occasionally heavy rainfall events lasting more than a day. During unusually intense El Niño events, a storm can enter Arizona with its circulation intact but with its severity downgraded from hurricane to tropical storm. This phenomenon is known to have happened three times during the 20th century: Hurricane Joanne on 6 October 1972; Hurricane Kathleen on 10 September 1976; and Hurricane Nora in September 1997 (Chenoweth and Landsea 2004). On a climatological time scale, moisture from dissipating tropical storms accounts for about one-half of the warm-season rainfall in Arizona.

Variations in Temperature and Precipitation

Besides variations on annual and seasonal time scales, Arizona has also experienced significant variations in both temperature and precipitation patterns on a decadal time scale. Table 3.1 summarizes decadal variations in temperature and precipitation patterns based on data obtained from Green and Sellers (1964) and the National Oceanic and Atmospheric Administration's Western Region Climate Center for 68 weather stations located throughout the state. The stations selected have complete precipitation records going back at least to 1953 and have undergone only minor relocations since their establishment. Thirty-one of the 68 stations also have complete records of maximum and minimum temperatures since 1953. Seven stations have reliable temperature records prior to 1943. They are the Casa Grande National Monument, Childs, Miami, Saint Johns, Snowflake, Williams, and the Winslow Airport. Precipitation records for Snowflake and Williams go back to 1883.

The average standard deviations for the maximum and minimum temperature and precipitation in the five decades between 1953 and 2002—that is, 0.6°F for the maximum temperature, 0.8°F for the minimum temperature, and 0.95 inch for precipitation (table 3.1)—are reasonable estimates of the true statewide standard deviations for all time periods.

There has been little consistent trend in the maximum temperature. None of the average anomalies in table 3.1 is significantly greater than zero. However, based on the "percent" data, the decade with the lowest average maximum temperature since 1953 was probably 1963 to 1972, while the decade with the highest value was 1993 to 2002. The increase in the maximum temperature between the two periods appears to be about 2°F, with increases of more than 3.0°F occurring only in Duncan (3.7°F) and Miami (3.2°F).

The data for the minimum temperature in table 3.1 are more revealing. Once again, the coldest of the five decades was apparently that from 1963 to 1972 and the warmest that from 1993 to 2002, with the increase in the minimum temperature between the two periods being about 2°F, the same as for the maximum temperature. Most of this increase in both the maximum and minimum temperatures since 1963 can be attributed to the "enhanced greenhouse effect," where increased carbon dioxide and water vapor in the atmosphere trap heat energy emitted by the Earth's surface. This phenomenon is probably the most significant indication of global warming in Arizona.

The maximum temperature in Phoenix increased by only 1.6°F from the 1963–1972 decade to the 1993–2002 decade, but the minimum temperature went up 6.9°F, the largest increase in the state. This increase in the minimum temperature is evidence of an urban heat-island effect, where night temperatures in large cities are increased by the heat storage in buildings and paved roadways and reduced by wind speed and evaporation. Of the other 30 stations used in this analysis, the minimum temperature increased by more than 4°F only at Sedona (4.5°F), Snowflake (4.2°F), and the Campbell Avenue Experimental Farm in Tucson (4.4°F). Data from the University of Arizona weather station dating back to 1892 were not used in this analysis because of recent changes in the location and elevation of this station. However, these data suggest the same results as shown by the data for Phoenix, with an increase of 6.1°F in the minimum temperature and a decrease of 0.5°F in the maximum temperature from the 1963–1972 decade to the 1993–2002 decade.

The data presented in table 3.1 also suggest that the minimum temperature—but not the maximum temperature—averaged significantly higher during the 1953–2002 period than it did prior to the early 1940s. Data from the University of Arizona are generally in accord with this trend. Although the maximum temperature for the 1953–2002 period averaged 1.4°F higher than for the period 1893–1942, the minimum temperature increased almost 5°F in the same time interval. Nighttime temperatures prior to 1943 apparently reached their lowest value in the low-lying desert areas of the state on the morning of 7 January 1913, when the temperature dropped to 5°F in Benson, 11°F in Gila Bend, 16°F in Phoenix, and 6°F in Tucson (Sellers et al. 1987). Temperatures as low as 0°F were reported at the lower elevations along the washes in the Tucson area. In January 1937, the Tucson temperature was below normal every day

Table 3.1 Decadal temperatures and precipitation summaries for selected weather stations[1]

	1883–92	1893–1902	1903–12	1913–22	1923–32
Maximum temperature					
N				2	5
Anomaly (°F)				0.6	0.2
SD (°F)					0.8
Percent				100	40
Minimum temperature					
N				2	5
Anomaly (°F)				−3.6	−2.9
SD (°F)					3.0
Percent				100	80
Precipitation					
N	2	2	4	11	21
Anomaly (in.)	−0.86	−3.71	1.35	2.04	0.60
SD (in.)			1.35	1.69	1.66
Percent	100	100	100	91	56

[1]All values are anomalies relative to the average for the period from 1953 to 2002. N, number of stations; SD, standard deviation; Percent, the percent of stations having anomalies of the same sign as the average anomaly.

of the month. The minimum was in the teens on 3 days and the twenties on 15 days. It is difficult to imagine such low readings with the present climate of the state. The most likely cause for most of the increase in the average minimum temperature in Arizona since the early 1940s is the rapid expansion in the population and the industrial development that occurred in the area during and following World War II. If this assumption is correct, it is more evidence of the effect of man on the climate.

The decadal average annual statewide precipitation between 1953 and 2002 has varied irregularly by almost 4 inches (table 3.1), with the driest interval from 1953 to 1962 and the wettest from 1983 to 1992 coinciding, respectively, with the periods of least and greatest El Niño activity. The decade from 1993 to 2002 was also dry, with

1933–42	1943–52	1953–62	1963–72	1973–82	1983–92	1993–2002
7	14	31	31	31	31	31
0.1	0.0	0.2	−0.7	−0.4	−0.3	1.1
0.7	0.6	0.6	0.5	0.7	0.7	0.7
71	68	63	90	73	71	97
7	14	31	51	31	31	31
−1.6	−0.8	−0.3	−1.1	−0.3	0.5	1.2
1.8	1.2	0.9	0.8	0.7	0.8	0.9
71	64	66	92	60	77	94
26	42	68	68	68	68	68
−0.09	−1.47	−1.69	0.09	0.14	2.21	−0.75
1.14	1.13	0.84	0.76	1.19	1.06	0.89
54	93	96	60	59	100	82

drought conditions at this time accentuated by higher average temperatures than during earlier periods. Also, the occurrence of the most recent dry period (1993–2002) immediately following perhaps the wettest decade on record (1983–1992) could have given many people the perception that Arizona was in the throes of a record-breaking drought. However, it was the wetness of the 1983–1992 decade that was more unusual than the aridity of the 1993–2002 decade.

On a decadal time scale, these and other data indicate that, prior to the 1950s, the state was generally dry from the beginning of record-keeping through 1904 and wet from 1905 to 1922. The drought at the turn of the 20th century ended with a bang in 1905, with the wettest year on record. Statewide precipitation that year averaged almost 26 inches. Other especially wet years were 1978 (22.07 inches), 1941 (20.93 inches), and 1965 (19.27 inches). Two of the four driest years on record occurred in the 1950s, with 6.70 inches of precipitation in 1956 and 7.90 inches in 1950. The other dry years were 2002, with 7.95 inches of precipitation, and 1989, with 8.53 inches.

Elevational Gradients

Elevation is the main integrator of precipitation and temperature and their respective variability across Arizona's diverse landscapes. The large differences in elevation produce a vegetation continuum reflecting the changes in precipitation and temperature from the low-elevation hot and dry deserts to the high-elevation cool and moist forests. These changes in vegetation were documented in the early publications of Shreve (1915), Pearson (1920a, 1920b), Sykes (1931), Hart (1937), Schwalen (1942), and Lull and Ellison (1950), who collectively reported that the annual precipitation in the southwestern United States increases from desert floors to mountaintops throughout the state at a rate of about 4 to 5 inches per 1000-foot increase in elevation. However, this elevation gradient can be too high in some situations. The increase in precipitation will usually be greater at higher elevations than at lower elevations during wet periods, enhancing the rate of increase of precipitation with elevation as a consequence. These early studies were based on data from the first half of the 20th century, a generally wet period throughout the state. A lower value of a 3- to 4-inch increase in annual precipitation per 1000-foot increase in elevation is probably more appropriate in the current climatic conditions. Based on recent data, total annual precipitation increases by about 20 inches a year going from Tucson at 2500 feet elevation to the top of nearby Mount Lemmon at 8000 feet, or by about 3.6 inches per 1000 feet.

The wide range of elevations found in Arizona, along with variable temperature and precipitation regimes, has led to an extremely diverse flora in the state. The elevation of a plant community or association amplifies the combined effects of variable temperature and precipitation regimes on the distribution and sustainability of that plant community (Swetnam and Betancourt 1998). This elevational effect is especially significant when it is considered within the context of prolonged periods of excessive precipitation or prolonged drought and long-term climate change.

4
Geologic and Landscape Evolution

JON E. SPENCER

Arizona's landscape, which has changed dramatically and repeatedly over geologic time, is the stage on which Arizona's natural environments have evolved. This landscape, in turn, is a consequence of Arizona's geologic and hydrologic histories. The rocks that form the geologic bedrock reveal an enormously long geologic history extending back for 1.8 billion years. Arizona's present landscape is largely the product of an active geologic history extending over the past 30 million years that encompasses less than 2 percent of the duration of its geologic history. Volcanic activity and accumulating displacements on earthquake faults radically altered the state's landscape during this more recent time—especially in southern and western Arizona. This changing landscape rerouted rivers and produced closed basins that trapped surface runoff to form lakes and playas. Regional uplift of the Colorado Plateau during the past 50 million years (the timing is poorly known) was followed by incision of the Grand Canyon and its tributaries, accelerated erosional removal of sediments from the plateau region, and sediment deposition in the low desert regions and in the Salton Trough.

This chapter on the state's geologic and landscape evolution begins with an outline of Arizona's three physiographic provinces, is followed by an overview of the geologic origin and evolution of Arizona's bedrock, and then focuses on its past 30 million years of geologic, hydrologic, and landscape evolution.

Physiography

Geologists divide Arizona into three physiographic provinces that resulted from the state's geologic history and extend beyond the borders of the state (fig. 4.1). The Colorado Plateau forms the north-eastern part of Arizona and is characterized by mesas, such as Black Mesa on the Navajo and Hopi Reservations, and the deeply incised and geologically young Grand Canyon and its tributaries. The rocks that make up the Colorado Plateau are mostly flat-lying to gently dipping sedimentary rocks, such as those that are so well exposed in the Grand Canyon, Monument Valley, and Canyon de Chelly. These sedimentary rocks are overlain in several areas by geologically young volcanic fields, most notably the San Francisco volcanic field in the Flagstaff area and the Springerville volcanic field between Show Low and Springerville.

The Transition Zone physiographic province is a northwest-trending belt between the Colorado Plateau to the northeast and the Basin and Range physiographic province to the southwest. Its geology is somewhat like that of the Colorado Plateau because significant parts of it are underlain by gently dipping sedimentary rocks (for example, in the Sedona area). The boundary between the Transition Zone and the Colorado Plateau is the Mogollon Rim—named after a feature north of Payson—that is generally located at the top of southwest-facing cliffs along the edge of the Colorado Plateau or is placed near the drainage divide within young volcanic rocks (Peirce 1984a). Elevations across the Transition Zone generally decrease northeast to southwest from 5000 to 8000 feet on the northeast to 2000 to 4000 feet on the southwest. Within much of this province, the sedimentary rock layers that make up much of the Colorado Plateau have been stripped off and the underlying old granitic and metamorphic rocks are widely exposed, such as those found in the area around Prescott and in the Mazatzal Mountains southwest of Payson. These rocks are part of the crystalline basement that also makes up the deepest part of the Grand Canyon.

The Basin and Range Province forms southern and western Arizona and extends southward far into Mexico and northward across

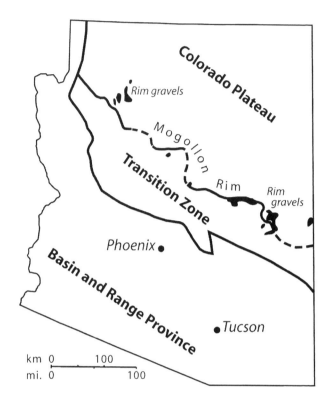

Figure 4.1. A simplified geologic map of Arizona showing its three major physiographic provinces, the Colorado Plateau, Transition Zone, and Basin and Range Province.

Nevada and western Utah to the Snake River Plain (Stewart 1978; Henry and Aranda-Gomez 1992), as shown in figure 4.2. Its characteristic form of small rugged mountain ranges and alluvial basins is the result of the last 30 million years of faulting and volcanism. The rocks exposed in these numerous mountain ranges are much like those of the Transition Zone in some areas, but in other areas they have been severely disrupted by younger faulting, igneous activity, and metamorphism. Most of Arizona's human population lives on

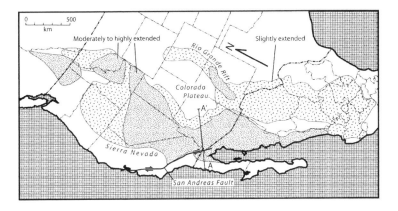

Figure 4.2. Southwestern North America, showing areas that were extended by fault movements during the past 50 million years. The Basin and Range Province includes the extended area south of central Idaho but not the Rio Grande Rift in New Mexico and Colorado.

the alluvial basins of the Basin and Range Province. The poorly consolidated sand and gravel that typically underlie these basins contain large groundwater reserves that have sustained much of Arizona's population and agricultural growth. Huge bodies of salt (halite) and gypsum buried in some of these basins were deposited in lakes, probably at a time when Arizona was wetter than it is currently.

Bedrock Geology

The Earth's crust beneath all of Arizona was initially created 1.8 to 1.6 billion years ago by geologic processes similar if not identical to currently occurring processes at convergent plate boundaries (Bowring and Karlstrom 1990; Eisele and Isachsen 2001). The modern island arcs of the western and northern Pacific Ocean, such as those represented by the Mariana and Aleutian Islands, are sites of volcanic activity produced where tectonic plates converge and one sinks into the deep mantle beneath the other plate—a process known as subduction. New crust is created in part by addition of volcanic rocks. Sediments are scraped off the sinking plate as well as some basaltic sea floor and added to the island arcs. Sediments are shed off the volcanic islands into the surrounding waters. Over geologic time, large land masses such as the Aleutian Peninsula were built by these processes. Enormous volumes of sediments can be buried to great depth and metamorphosed into hard rock. Magmas ascending from deep in the Earth can melt and assimilate such crustal rocks and produce granitic rocks. Continental crust such as that beneath Arizona is thought to have formed over about 200 million years by such processes. Many granitic rock bodies in the state were produced during this time, and much schist, such as the Pinal Schist in south-eastern Arizona, was created by metamorphism of marine sedimentary rocks.

A poorly understood period of magmatism that occurred across North America about 1.4 billion years ago added enormous bodies of granite to Arizona (Anderson 1989). The Oracle Granite north of Tucson, Ruin Granite north of Globe, and Carefree Granite northeast of Paradise Valley near metropolitan Phoenix were all produced at this time. These granites tend to have few quartz veins but are high in iron and so contain much black biotite and magnetite. Some of the black magnetite sands notable in some desert washes were derived from this type of granite.

The pace of geologic evolution in Arizona slowed dramatically after these granites were intruded. For the next billion years or so, the area was eroded to a relatively flat surface that was intermittently blanketed by thin layers of sediment. The oldest such veneer of sediments, known as the Apache Group (Wrucke 1989), consists of several hundred feet of what is currently quartzite, slatey siltstone, and a relative of limestone known as dolomite in which (ideally) every other calcium atom is replaced with a magnesium atom. At about 1.1 billion years ago, the Apache Group was intruded by horizontal sheets of basaltic magma that currently appear as dark green to black rock layers tens to hundreds of feet thick. Both the Apache Group and intruding basaltic layers are well exposed over much of Gila County and form some of the large road cuts along State Route 77 in the Salt River Canyon.

A second veneer of limestone, dolomite, sandstone, and siltstone was deposited across Arizona during the Paleozoic Era (544 to 245 million years ago; Dickinson 1989), as illustrated in figure 4.3. These sediments form most of the walls of the Grand Canyon, with older rocks forming the lower inner gorge. The Kaibab Limestone at the top of this sequence forms the rim of the Grand Canyon and underlies much of the surrounding area, including the extensive and mostly low-relief Coconino Plateau south of the Grand Canyon (fig. 4.4). Overlying sedimentary rocks from the Mesozoic Era (245 to 65 million years ago) consist largely of sandstone and siltstone that form much of the bedrock geology on the Navajo and Hopi Reservations, including many of its scenic cliff faces and other outcrops. These cliff-forming rock units include the extensive sand-dune deposits of the Jurassic Navajo Sandstone and the Cretaceous coal-bearing sandstone and siltstone on Black Mesa. Explosive volcanism in southern Arizona during the Jurassic Period (208 to 146 million years ago) produced extensive volcanic rocks, some of which overlie quartzite derived from windblown sand, for example, in the northern Santa Rita Mountains south of Tucson.

The Mesozoic Era ended with a meteor impact that caused global extinction of numerous flora and fauna. At this time, southern and western Arizona were in the middle of a period of volcanism and faulting that was more severe than anything in its previous billion years of geologic history. Renewed and vigorous subduction of the Pacific Ocean sea floor beneath southwestern North America caused both magmatism and faulting-related mountain building. This episode of magmatism, faulting, and uplift known as the Laramide orogeny (an *orogeny* is a period of mountain building) occurred between about 80 and 50 million years ago (Coney and Reynolds 1977; Dickinson and Snyder 1978; Krantz 1989). Volcanoes were probably widespread during the Laramide orogeny, but because the volcanoes were generally built on a mountain range, they were largely or completely eroded away over the following tens of millions of years, exposing the crystallized magma chambers that had been beneath the volcanoes. These magma chambers are currently represented by granitic rocks that underlie many areas of southern and western Arizona (Richard et al. 2000). Such granitic rocks are widely

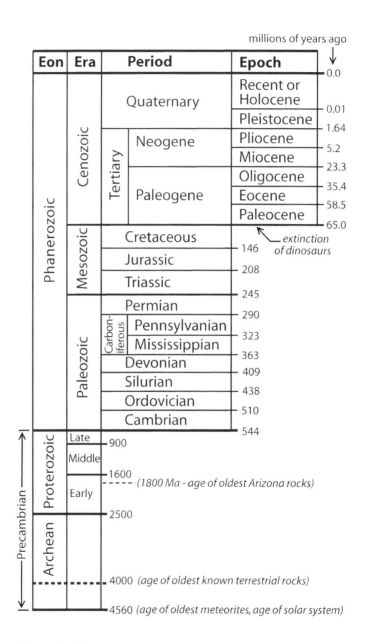

Figure 4.3. The geologic time scale.

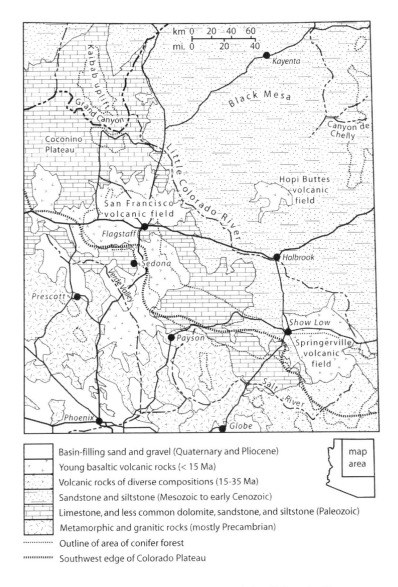

km 0 20 40 60
mi. 0   20   40

Kaibab uplift

Grand Canyon

Coconino Plateau

Black Mesa

Kayenta

Canyon de Chelly

Little Colorado River

Hopi Buttes volcanic field

San Francisco volcanic field

Flagstaff

Sedona

Verde Valley

Prescott

Holbrook

Show Low

Payson

Springerville volcanic field

Salt River

Phoenix

Globe

Basin-filling sand and gravel (Quaternary and Pliocene)
Young basaltic volcanic rocks (< 15 Ma)
Volcanic rocks of diverse compositions (15–35 Ma)
Sandstone and siltstone (Mesozoic to early Cenozoic)
Limestone, and less common dolomite, sandstone, and siltstone (Paleozoic)
Metamorphic and granitic rocks (mostly Precambrian)
.... Outline of area of conifer forest
.... Southwest edge of Colorado Plateau

map area

Figure 4.4. A simplified geologic map of the Colorado Plateau region of northeastern Arizona. Outlined in the figure are the coniferous forests that extend over a diversity of rock types within and on the edge of the Colorado Plateau. Ma, millions of years ago.

exposed in the mountain ranges around Tucson, the Picacho and Sacaton Mountains near Casa Grande, and the White Tank Mountains west of Phoenix.

Intrusion of granitic magmas during the Laramide orogeny produced Arizona's abundant copper deposits (Titley 1995). The magmas responsible for producing these and related deposits contained abundant sulfur and metals that were deposited as sulfide minerals, including the iron-sulfide mineral pyrite (FeS_2) and the copper and iron-sulfide mineral chalcopyrite ($CuFeS_2$). As various sulfide mineral deposits were uncovered by erosion, the sulfide minerals were oxidized by exposure to shallow groundwater containing dissolved oxygen and to atmospheric oxygen (Titley and Marozas 1995). This process continues today as Arizona's numerous sulfur-containing mineral deposits are gradually uncovered by erosion. Oxidation of pyrite in the presence of water produces iron sulfate ($FeSO_4$) and sulfuric acid (H_2SO_4). Acid liberated by sulfide oxidation locally decreases the pH of groundwater and surface water, a natural process that can be enhanced where sulfide minerals are uncovered by mining and ultimately reduced where the sulfides are removed by mining and taken to a smelter. (Large smelters produce thousands of railroad tank cars of sulfuric acid that are taken away for industrial uses.) Rocks and derived talus, gravel, and sand in areas of sulfide mineralization are commonly yellowish orange, pale orange, or reddish brown because of the presence of limonite—a mixture of hydrous iron oxides and sulfates—deposited from water that was mildly acidic and carrying iron.

Middle Cenozoic Volcanism and Crustal Extension

Drastic geologic changes between about 30 and 15 million years ago produced a rough approximation of Arizona's modern topography. Voluminous and explosive volcanism built the volcanic rock masses that currently form much of the Chiricahua, Galiuro, Superstition, and Big Horn Mountains; much of the area between Highways I-10 and I-8 west of Gila Bend; the mountains between Gila Bend and northern Sonora; the Black Mountains of western Mohave County;

and almost all of Greenlee County and adjacent areas in the eastern part of the San Carlos Apache Reservation (Shafiqullah et al. 1980; Sheridan 1984; Spencer et al. 1995). The most explosive eruptions during this period produced many tens to hundreds of cubic miles of hot ash that each catastrophically buried tens to thousands of square miles with tens to hundreds of feet of what is currently hard volcanic rock. The imposing cliffs on the southwestern side of the Superstition Mountains are composed of volcanic ash from one of these violent eruptions 18 million years ago. Severe faulting and erosion have largely obliterated the volcanic landforms produced by these eruptions, and the original form of the volcanoes and calderas can be pieced together only by careful geologic field mapping.

Normal faults are those faults that dip moderately to gently into the Earth and in which rocks above the fault moved generally down the dip of the fault. Movement on these faults results in horizontal extension of the Earth's crust. Such normal faults largely destroyed the mountain belt across Arizona that was built up by the earlier Laramide orogeny. Extension of the Earth's crust accommodated by northeast–southwest movement on normal faults approximately doubled the width of the land surface and reduced the crustal thickness by one-half (Spencer and Reynolds 1989), as shown in figure 4.5. This mountain belt was reduced to moderate to low elevations, and the modern topography of basins and ranges developed. From a geophysical perspective, the potential energy stored in the elevated topography of the mountain chain was released as elevations decreased and the crust extended horizontally by faulting (Jones et al. 1998). In somewhat informal usage, the mountain range is said to have collapsed. Multiple parallel normal faults moving at the same time allowed fault blocks to tilt over like a set of books on a shelf that all spread out as they fall over. This type of faulting was so severe in some areas that the volcanic rocks and their interlayered sedimentary rocks are tilted up to 90° and are even locally overturned, for example, in the Vulture Mountains south of Wickenburg.

Some normal faults had a dip of less than about 30° and accommodated displacement of tens of miles. Such large displacement resulted in uplift and exposure of rocks that formerly had been in

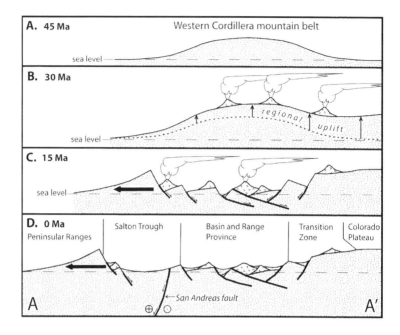

Figure 4.5. Schematic cross-sections showing the evolution of the Basin and Range Province and the Salton Trough in Arizona and Baja California Norte (see fig. 4.2 for location). The older period of faulting and magmatism produced a mountain belt (A) that was later raised by regional uplift during a period of voluminous magmatism (B). This uplift event also elevated the Colorado Plateau. Uplift and magmatism were accompanied and later followed by extension and collapse of the western Cordillera mountain belt (C). On cross-section (D), dominantly horizontal movement on the San Andreas fault is indicated by two circular symbols on opposite sides of the fault. Because this is a cross-section view, rocks on one side of the San Andreas Fault are moving toward the viewer (circle with dot representing an arrow tip), and those on the other side are moving away (circle with cross representing an arrow tail).

the middle of the Earth's crust, where rocks are normally too hot and soft to fracture and slip during earthquakes. Instead, such rocks deform by flowing like hot and soft plastic. These rocks currently appear as layered gneiss, like that on the southern flank of the Santa Catalina Mountains, or as layered granite, like that in the eastern half of the South Mountains south of Phoenix. The eastern parts of the Harquahala, Harcuvar, Buckskin, and Rawhide Mountains in western Arizona and the southwestern flank of the Picacho, Tortolita, and Rincon Mountains near Tucson consist of such rocks. These areas of plastically deformed rocks originating from the middle crust and carried to the Earth's surface by normal faulting are known as metamorphic core complexes (Davis 1980; Davis et al. 1986; Spencer and Reynolds 1989).

Late Cenozoic Volcanism and Crustal Extension

The voluminous volcanism, extreme crustal extension, and normal faulting that so drastically modified the Arizona landscape between 30 and 15 million years ago were followed by a period of volcanism and normal faulting that was less severe in its consequences and different in some of its qualities. While previous volcanism was explosive in nature and associated with subduction beneath the adjacent continental margin, beginning 15 million years ago volcanism was dominantly basaltic in character and unrelated to subduction (Spencer and Reynolds 1989). Basaltic lavas flowed great distances but rarely produced explosive eruptions or large volcanoes (Sheridan 1984). The extensive and dominantly basaltic lava fields around Flagstaff and east and south of the Verde Valley were produced during this time, as was the Hopi Buttes volcanic field north of Holbrook (fig. 4.4). Volcanism around the San Francisco Peaks has been substantial in the past 2 million years and occurred within the past 2000 years at Sunset Crater near Flagstaff (Wolfe 1984). Basaltic lava fields that erupted during the past 2 to 3 million years include the Sentinel volcanic field along Highway I-8 west of Gila Bend, the San Bernardino volcanic field northeast of Douglas, and numerous lava flows north of the western Grand Canyon—some of which spilled into the canyon and temporarily dammed it (Lynch 1989;

Hamblin 1994). The Springerville volcanic field east of Show Low was also produced during this period, and many volcanic landforms such as cinder cones and volcanic craters remain visible because the volcanism is geologically recent.

Normal faults that were active during the past 15 million years generally dip more steeply than normal faults of the earlier period. Modern basins and ranges are directly related to movement on these faults in some cases. About 2 miles of slip on the Pirate fault, which is exposed along the western foot of the Santa Catalina Mountains, has created Oro Valley and uplifted the western Santa Catalinas. Similar range-bounding normal faults—many of which are buried and known only from geophysical investigations—dip beneath basins, for example, along the northeastern flank of the Pinaleño Mountains and the eastern flank of the Baboquivari Mountains. Some of these faults have been active in the past million years, such as the faults at the eastern foot of the southeastern Pinaleño Mountains and the northwestern flank of the Santa Rita Mountains. The active Pitaicachi fault in northeastern Sonora, Mexico, southeast of Douglas, produced an earthquake in 1887 that resulted in dozens of fatalities in Sonora and extensive damage to buildings. Numerous rock falls caused by the earthquake were followed by fires in many mountain ranges in Sonora and southeastern Arizona, possibly because of frictional heating during collisions between large, fast-moving rocks and stationary rocks (DuBois and Smith 1980). Many active normal faults northwest of Flagstaff and along the Wasatch Front extending from the northwestern corner of Arizona northward to Salt Lake City and beyond are also active and still contributing to uplift of range fronts and occasional earthquakes (Pearthree and Bausch 1999).

Uplift of the Colorado Plateau

The area that is currently the Colorado Plateau had been near sea level before the Laramide orogeny, as indicated by marine strata deposited in northwestern Arizona shortly before 80 million years ago (Nations 1989). After the Laramide orogeny, streams carried sediments from southern and western Arizona onto what is currently the Colorado Plateau and into New Mexico (see the "Rim Gravels" in fig. 4.1; Cather and Johnson 1984; Young 2001). The regional topography of the southwestern United States was generally the reverse of what it is currently, with a low area forming what is now the Colorado Plateau and a high area to the southwest.

The timing of uplift of the Colorado Plateau from near sea level to its present elevation of about 1.5 miles has been a controversial and unresolved issue among geologists. One school of thought is that the uplift has been fairly recent and broadly synchronous with late Cenozoic basaltic volcanism and normal faulting. The canyons and mesas of the plateau are geomorphically youthful features, and, therefore, it has been argued that the Colorado River cut the Grand Canyon in the past few million years as the plateau was uplifted (Lucchitta 1979). Others have argued that the Colorado Plateau was uplifted earlier, as indicated by paleofloral analysis that suggests elevations in southern Colorado were similar to modern elevations 35 million years ago (Wolfe et al. 1998).

Late Cenozoic Arizona: Land of Lakes

Over geologic time, volcanism and faulting can produce hydrologically closed basins by damming rivers with lava flows or with uplifts produced by fault movements. Integration of such closed basins with river systems and termination of lake or playa environments occurs due to sediment filling, lake overflow, and outlet incision that could be aided by erosional expansion of nearby unconnected drainage systems. Both outlet incision and sediment infilling are slower in arid environments than in areas with moderate or abundant precipitation. Termination of volcanism and faulting is followed over geologic time by drainage integration.

Lakes in the southwestern United States are typically ephemeral and occur primarily as playas. The Death Valley and eastern Mojave Desert areas of California are subject to active faulting and contain numerous playas. Active faulting, low elevation, and conditions of aridity have all contributed to the creation and preservation of these playas. The numerous playas and lakes in the Great Basin of Nevada and western Utah are a consequence of the movement on numerous normal faults over the past few million years. Continued normal faulting on the flanks of the Great Basin—specifically on the eastern flank of the Sierra Nevada and west of the Wasatch Front in Utah—contributes to the formation and maintenance of lakes and playas such as the Great Salt Lake. Arizona was subjected to a similar environment of normal faulting. However, such faulting and basin formation largely ended during the past 5 to 10 million years, and basin filling, spillover, and integration with river systems has terminated the closed basin hydrologic conditions in almost all parts of the state. Wilcox playa in southeastern Arizona and Red Lake playa north of Kingman are the two remaining survivors of Arizona's geologically recent past as a land of lakes.

The numerous lakes and playas that occupied Arizona's landscape during the last 15 million years left voluminous salt deposits that are currently buried beneath basins. The buried Red Lake salt body near Kingman contains approximately 50 cubic miles of halite ($NaCl$) with an anhydrite ($CaSO_4$) cap (Faulds et al. 1997). The buried Luke salt body on the western side of Phoenix is several thousand feet thick (Peirce 1984b). Incised lake beds consisting of sandstone, siltstone, and in some cases limestone are widely exposed in the Verde Valley, the Big Sandy Valley southeast of Kingman, the lower San Pedro Valley, and from the Safford area south to Highway I-10 (Scarborough 1989). Deep drilling operations in the alluvial basins of southern and western Arizona commonly encounter clay, halite, anhydrite, and gypsum ($CaSO_4 \cdot 2H_2O$), all of which were deposited in the lakes or playas of the state's geologically recent past.

The Colorado River

The Hualapai Limestone was deposited between 10 and 6 million years ago in a lake in the eastern Lake Mead area. Most of it is exposed on the southeastern side of Lake Mead, where it forms Grapevine Mesa. The lower and older part of the Hualapai Limestone was deposited in the central portion of Grand Wash Trough, which is the north–south trending valley that the Colorado River enters where it exits the Colorado Plateau and western Grand Canyon. Sand, gravel, and conglomerate were deposited by alluvial fans on both sides of Grand Wash Trough, with the limestone in the middle (Lucchitta 1987). A large river like the Colorado River could not have entered Grand Wash Trough at this time without rapidly filling the basin with sediments, spilling over at its lowest overflow point, incising an outflow channel, and emptying the lake. This appears to be what happened about 5 million years ago and, furthermore, it seems clear that the Colorado River did not arrive in Grand Wash Trough until that time. The largest single unsolved problem in understanding the hydrologic evolution of the southwestern United States during the past 25 million years concerns the fate of all of the water that flowed off the western side of the Colorado Rocky Mountains and southern flank of the Uinta Mountains before the modern Colorado River entered the Lake Mead area. If it left a currently abandoned river valley, that valley has not been found. If the water flowed into a lake and evaporated, the considerable sedimentary deposits it would have left have not been found. The Bidahochi Formation in east-central Arizona north of Holbrook includes siltstone that could have been deposited in a lake, but its elevation is problematically high (Ort et al. 1998).

Abrupt arrival of Colorado River water to Grand Wash Trough led to rapid filling of a lake that would have flooded the lower Virgin River area and spilled over into the vicinity of the Las Vegas area. Limestone and gypsum at Frenchman Mountain east of Las Vegas record this brief inundation (Castor and Faulds 2001). Spillover into Mohave Valley and southward through a sequence of basins left from earlier normal faulting led to deposition of flood deposits marking initial spillover (House et al. 2002). Silty limestone and siltstone of the Bouse Formation record the presence of lake water along the Lower Colorado River Valley from north of Bullhead City south to the Chocolate Mountains south of Blythe (Metzger and Loeltz 1973; Metzger et al. 1973; Spencer and Patchett 1997). In the Parker-Blythe area, a large lake formed for probably several thousand to tens of thousands of years before spillover and incision of an outflow channel north of Yuma. This lake was apparently sufficiently salty that marine fauna transported inadvertently by birds colonized the lake and left fossils and shells, which appear to indicate that the Bouse Formation was deposited in an estuary (Smith 1970; Todd

1976). According to the estuary interpretation, the maximum elevation of the Bouse Formation records ancient sea level of an enormous area that subsequently was raised by tectonic forces to present elevations (Lucchitta 1979). This interpretation, however, is incompatible with both the strontium isotopic composition (an indicator of seawater or runoff) of the Bouse Formation and its association with flood deposits derived from the north. It is more likely that the Bouse Formation was deposited in lakes that represent the first arrival of Colorado River water to a chain of closed basins (Spencer and Patchett 1997; House et al. 2002).

The Last 5 Million Years

Increased hydrologic maturity associated with the integration of drainages into the Colorado, Gila, and Bill Williams River systems has occurred over the last 5 million years. The dramatic and geologically youthful landscape of the Colorado Plateau in northern Arizona is largely the result of incision and sediment stripping of the landscape during this period (fig. 4.6). Aggradation continues to fill basins inherited from the earlier period of faulting and volcanism, primarily in the Basin and Range Province. The Santa Cruz and Rillito Rivers are incised in the Tucson area, merge, and continue northward through Marana, where they exit a confined channel and spread out over the wide valley floor during floods. Wide and shallow river flow between Marana and the Gila River, especially characteristic of the time prior to human modification, so reduced water velocity that all but the finest sediments settled out of the moving water. Stream incision created river channels in other areas and, in upland areas, carved ridges and hills from previously deposited alluvial fans and lake sediments. The hilly country around Sonoita in southeastern Arizona is a good example of this type of incised landscape (Menges and Pearthree 1989).

Incision of the Salt, Verde, Agua Fria, and Hassayampa River channels near Phoenix is more pronounced upstream near the points where these rivers exit bedrock canyons and decreases downstream. This incision has led to the inference that northern areas of the state,

Figure 4.6. The Grand Canyon is a notable feature of the geologically youthful landscape of the Colorado Plateau in northern Arizona.

including the Colorado Plateau, have undergone uplift, while stream gradients have remained unchanged (Péwé 1978). Other areas of significant Quaternary incision, such as where the Cañada del Oro exits the Santa Catalina Mountains and begins its southward flow through Oro Valley, reveal no evidence of local uplift. Increasing incision of landscapes in the past 2 to 3 million years has been linked to climate change in which ice ages alternate with warmer periods, such as that of the past 10,000 years. This Quaternary climatic regime is thought to produce alternating periods of bedrock disaggregation and sediment removal, resulting in more effective incision and erosion than that associated with a more static and earlier climate regime (Peizhen et al. 2001).

5
Soil and Water Resources

LEONARD F. DeBANO, DANIEL G. NEARY, and PETER F. FFOLLIOTT

Soil and water form inseparable resources throughout Arizona. Their roles consist of many physical, chemical, hydrologic, and biological properties and processes. This interactive and highly complex soil-water system supplies air, water, nutrients, and mechanical support for the sustenance of plants in all of the state's plant communities. Hydrologic processes within the soil-water system also provide on- and offsite pathways for transferring incoming precipitation. Precipitation either flows over the soil surface or infiltrates in the soil mantle, where it eventually drains into stream channels and is transported downstream in larger stream systems and rivers. The quantity, quality, and timing of streamflow leaving upland watersheds and changes in these parameters during its flow to larger river basins can affect the benefits derived from the streamflow by downstream users.

Soils

Soils consist of minerals, organic matter, air, and water components that support a wide range of living organisms. The mineral component includes unaltered minerals that are part of the parent rock and secondary minerals formed by the recombination of substances produced during weathering (Hendricks 1985). Small particles of organic matter (humus) help hold mineral soil particles together to form aggregates and contribute to soil structure. The resulting soil structure provides pore space that allows air, water, plant nutrients and roots, and small organisms to move throughout the soil.

The minerals and organic matter found in soils are arranged in distinct horizontal layers, or soil horizons, that are largely parallel to the soil surface. The soil profile is the collective vertical arrangement of these horizons. The number of horizons in a soil profile can vary from one to several, depending mainly on the degree of rock weathering and soil formation. Soil profiles with many horizons are considered to be older and more developed compared to those with fewer horizons. Soil horizons are also used as diagnostic criteria for classifying some soils.

Soils are created as the end product of physical, chemical, and biological processes operating over long periods of time (Singer and Munns 1996). They are formed through continual interactions between the underlying geologic system (see chapter 4) and the topographic (slope, aspect, and altitude), climatic (temperature, air, and water), and biotic (faunal and floral) components of the natural environment. Resistance of rocks to weathering, inherent chemical composition, regional distribution, and topographic location all contribute to the properties of soils that develop within a particular geologic setting. Microclimate also plays a role in defining the soils found in Arizona's natural environments.

Soil Classification

Classification is fundamental to the understanding of soils and their relationship to water, vegetation, and other aspects of the natural environment. The classification system outlined in *Keys to Soil Taxonomy* prepared by the Soil Survey Staff (1998) is used in Arizona and elsewhere in the United States. This classification system focuses on measurable soil properties rather than the soil-forming processes that were a part of earlier classification systems. These properties are soil depth, texture, structure, moisture, temperature, clay mineralogy, cation exchange capacity, base saturation, organic matter content, and salt content. Measurement of these properties forms the basis for classifying soils into the hierarchical taxonomic categories of order, suborder, great group, subgroup, family, and series. While only orders and subgroups are considered in this chapter, a more detailed discussion of the other categories is presented in Hendricks (1985) and other references on Arizona's soils.

Five major soil orders, namely the Alfisols, Aridisols, Entisols,

Table 5.1   Soil subgroups in combinations of five soil orders and seven climatic regimes in Arizona

Temperature-moisture	Alfisol	Aridisol	Entisol	Mollisol	Vertisol	Total
Hypothermic Arid		17	4			21
MAT = 72°F or greater						
MAP = 10 in. or less						
Thermic Arid		6	5			11
MAT = 59–72°F						
MAP = 5–10 in.						
Thermic Semiarid	1	21	8	7	1	38
MAT = 59–73°F						
MAP = 10–18 in.						
Mesic Arid		1	12			13
MAT = 47–59°F						
MAP = 6–10 in.						
Mesic Semiarid	1	9	3	10	1	24
MAT = 47–59°F						
MAP = 10–16 in.						
Mesic Subhumid	7	1	1	7		16
MAT = 47–59°F						
MAP = 16 in. or greater						
Frigid Subhumid	7		5	6		18
MAT = 47°F or less						
MAP = 16 in. or greater						
Total for the subgroups	16	55	38	30	2	141

Source: Adapted from Hendricks 1985.

Notes: MAT, mean annual temperature at a depth of 20 inches or at the soil-rock interface in shallow soils; MAP, mean annual precipitation.

Mollisols, and Vertisols, reflect the wide range of soil-forming conditions occurring in the state. These soil orders contain 141 soil subgroups. A tabulation of the 141 soil subgroups formed by the combination of seven climatic (soil temperature–annual precipitation) regimes with the five soil orders provide a framework for discussing the soil-vegetation relationships found throughout Arizona (table 5.1).

Soil Orders

The Aridisols are the most common soil order in Arizona, where they make up about 39 percent of the 141 subgroups (table 5.1). Aridisols have light-colored surface layers, low amounts of organic matter, and at least one diagnostic subhorizon. Calcium carbonate is usually found in some or all parts of this soil order. Aridisols have developed in regions of Arizona with arid and semiarid climates.

The next most common soil orders are the Entisols, Mollisols, and Alfisols that contain 27, 21, and 12 percent of the 141 subgroups, respectively. Entisols can have thin surface horizons with some organic matter, but the parent material has not been altered enough to produce subsurface horizon development. They are found throughout all of the climate zones of Arizona. Entisols are considered relatively young soils because of the lack of subsurface horizon development. Mollisols have thick and dark-colored surface horizons. These soils are found in semiarid and subhumid climate zones at higher elevation on landscapes covered by grasses, grass-tree mixtures, or mostly trees in some parts of the state. Alfisols are soils with light-colored surface layers and clayey subsurface horizons. They occur at higher elevations in the state. Alfisols are relatively old soils, some more than 10,000 years old.

Vertisols are relatively rare in Arizona, being represented by only two subgroups, or about 1 percent of the 141 subgroups. Vertisols are clayey soils that are associated primarily with volcanic (basalt) rocks. They are composed dominantly of montmorillonite that undergoes considerable shrinking and swelling during drying and wetting cycles in the year. Vertisols are found mostly in the semiarid and subhumid climate zones of the state.

Inceptisols are seldom found in Arizona and therefore are not included among the subgroups tabulated in table 5.1. Inceptisols are relatively young soils that lack horizons of illuvial clay accumulations and occur only in subhumid regions. They have weak to moderate horizonation that evolves mostly from slight to moderate alteration of the parent material.

Soils and Climatic Regimes

Hyperthermic Arid soils cover about 27 percent of Arizona's land area. Subgroups found in this climatic (soil temperature–annual precipitation) regime mainly belong to the Aridisol soil group, with only a few Entisols (table 5.1). These soils are found at lower elevations in the Sonoran and Mojave Deserts in western and southwestern Arizona. The topography of these areas is gently to steeply sloping landscapes underlain by granitic and sedimentary rocks at intermediate elevations and by river alluvium along the floodplains of major lower elevation rivers, such as the Gila and Colorado. The soils are deep to moderately deep. Soil texture varies from fine-textured clay and silt material to larger gravels, which are frequently coated with lime.

Thermic Arid soils are found on about 8 percent of the state's land area. Subgroups developing under this climatic regime are almost equally split between the Aridisol and Entisol soil orders (table 5.1). Thermic Arid soils exist in parts of the Sonoran and Mojave Deserts at low to intermediate elevations in southern and north-central Arizona. These soils develop on volcanic materials at intermediate elevations and on river alluvium at lower elevations. Thermic Arid soils are deep on nearly level to gently sloping topography and shallow on hills and mountains. The topography, location, and parent materials of Thermic Arid soils are similar to the Hyperthermic Arid soils, although the mean annual soil temperatures and annual precipitation are less.

Thermic Semiarid soils cover about 20 percent of Arizona. The main subgroups in this climatic regime are mostly Aridisols, with lesser numbers of Entisols, Mollisols, and, only rarely, Alfisols and Vertisols (table 5.1). Thermic Semiarid soils are found in the Chihuahuan and Sonoran Deserts and support desert grassland vegetation at lower to intermediate elevations in southern and western Arizona. These soils are commonly derived from alluvium along lower elevation stream channels and volcanic and granitic rocks at intermediate elevations. Thermic Semiarid soils are well drained, distinctly stratified, and possess coarse to fine textures. These soils can be shallow to deep. They receive more precipitation than the closely related Thermic Arid soils.

Mesic Arid soils occur on about 12 percent of Arizona's land area. Subgroups in this climatic regime are primarily Entisols (table 5.1). The subgroups associated with this climatic regime are located mainly on the Colorado Plateau in the north-central part of the state, where they support Great Basin desertscrub. These soils occur on topographies ranging from eroded uplands and plateaus to nearly level floodplains along the Little Colorado River. The parent material of Mesic Arid soils is sedimentary shales and sandstones. These soils are shallow in the upland areas and deeper in the alluvial material deposited near the rivers. The most scenic of the Mesic Arid soils are those in the Painted Desert, Monument Valley, and Petrified Forest National Monument.

Mesic Semiarid soils occupy about 18 percent of Arizona's land surface. Subgroups formed in this climatic regime are composed primarily of Mollisols and Aridisols, with lesser numbers of Entisols, Alfisols, and Vertisols (table 5.1). Mesic Semiarid soils are located at intermediate elevations in northwestern and central Arizona, where they support plains grasslands and pinyon-juniper woodlands. These soils have developed on a wide variety of landscapes, including steep hills, mesa tops, mountains, rock outcrops, undulating hills, and plains, where they are derived from calcareous alluvium, sandstone, and shale. The depth to the calcareous substrate of Mesic Semiarid soils varies according to the degree of weathering and amount of precipitation (Ford et al. 2004). The soils are shallow on the rock outcrops of steep hills, mountains, and mesas and deeper on the floodplains and undulating to steep valley slopes and plains.

Mesic Subhumid soils cover 7 percent of the state. Subgroups in this climatic regime belong to the Alfisol and Mollisol soil orders, with only a few soils being Aridisols and Entisols (table 5.1). Mesic Subhumid soils occur in central Arizona below the Mogollon Rim and in southeastern Arizona at intermediate elevations, where they support interior chaparral communities and oak woodlands. Mesic Subhumid soils develop on rock outcrops on gently sloping to very steep igneous and sedimentary hills and mountains. The soils are mostly shallow and moderately coarse to moderately fine textured, but frequently contain large amounts of gravel and cobbles.

Frigid Subhumid soils also cover 7 percent of Arizona. Subgroups are almost evenly split among Alfisols, Entisols, and Mollisols (table 5.1). These soils are found mainly at high elevations on the Colorado Plateau and some of the higher mountains in the Basin and Range Province north of the Mogollon Rim. Frigid Subhumid soils support forests, mountain grasslands, and alpine tundra. The parent material is volcanic rhyolite and dacite, cobbly colluvium, basalt cinders and ash, and sandstones. The soils are moderately deep and medium to moderately fine textured and are found on moderately sloping to very steep mountain topography.

Soils in riparian (streamside) corridors (see chapter 6) are often of interest to people because of the high resource values associated with these streambank ecosystems. Riparian soils are widely distributed among the different soil groups and climatic regimes described above. Higher elevation soils can develop from both consolidated and unconsolidated alluvial sediments derived from a wide range of granitic, metamorphic, and sedimentary bedrock materials found in the surrounding uplands (Hendricks 1985). The depth of soils in riparian corridors varies widely, depending largely on the topographic setting and parent materials. The alluvial soils commonly found in lower elevation riparian corridors are subject to frequent flooding, are generally characterized by a wide range of textures, and can be sharply stratified, depending on the sediment depositional patterns occurring at the location. Riparian soils tend to be shallow, rocky, and gravelly at the higher elevations of the mountainous areas as compared to soils on the floodplains at lower elevations, where the recent depositions tend to be more uniform within individual horizontal strata, although they all have little profile development (Johnson et al. 1985; Medina 1986; Baker et al. 2004).

Water

The hydrologic processes regulating water and its disposition in Arizona are almost as varied as the state's soils and vegetation. Onsite processes of infiltration, percolation, surface runoff, and soil erosion are affected by diverse climatic patterns; complex geologic formations; and the physical, chemical, and biological properties of the soils on watershed landscapes. As water leaves upland watersheds, the offsite processes of streamflow, sedimentation, and water-quality determination come into play. A brief discussion of both sets of processes follows.

Infiltration, Percolation, and Surface Runoff

Precipitation that reaches the ground forms puddles on the soil surface, flows over the soil surface, or moves into the soil. Water that flows as surface runoff over the soil can lead to soil erosion. Water that enters the soil surface (infiltration) moves either downward (percolation) to groundwater aquifers or downward and laterally to a stream channel. Movement of water arriving at the soil surface into either surface or subsurface flow affects the timing and amount of streamflow that occurs. Water flowing over a soil reaches a stream channel in a shorter period of time than that flowing through the soil body.

Soil Erosion

The natural process of soil erosion occurs on all landscapes at different rates and scales. The physical process of this erosion consists of dislodgement and transport of soil particles by water and wind. The magnitude of soil erosion is affected by the climate (weather patterns), geology, soil characteristics, vegetative cover, and land use on a site. Peak soil erosion rates generally occur at the interface between desert grasslands and nearby shrub or woodland plant communities, where precipitation intensities are high enough to cause erosion but there is insufficient vegetation to protect against it (Brooks et al. 2003; Neary et al. 2005). Rates of erosion are accelerated when landscapes are denuded by disturbances such as wildfire, regardless of the precipitation regime. Forms of soil erosion on the upland watersheds of Arizona are surface erosion, gully erosion, and occasionally soil mass movement. Stream channel erosion also occurs as a result of water flowing in the channel.

Streamflow

Streamflow originating on high-elevation watersheds in Arizona flows through intermingling soils of volcanic, sedimentary, and granitic origin, whereas low-elevation streams flow mostly through alluvial deposits. Streamflow is generated when surface runoff (overland flow), baseflow, or combinations of the two flow pathways reach downslope stream channels. The state's streamflow regimes are perennial, intermittent, or ephemeral (fig. 5.1). The limited perennial streamflows tend to originate at the higher elevations, whereas ephemeral and intermittent streamflow origins are more common in lower desert environments.

Three main types of streamflow-generating mechanisms occur in Arizona: spring snowmelt-runoff events; low-intensity, long-duration winter rains; and high-intensity, short-duration convectional rainstorms during the summer months (Lopes and Ffolliott 1993; Lopes et al. 2001). Streamflow events produced by snowmelt-runoff can continue for several days or weeks during the late winter and early spring months. Up to 70 percent of the water flowing into the reservoirs on the Salt and Verde Rivers in central Arizona on an annual basis is produced by streamflow originating as snowmelt-runoff in late winter and early spring and frontal rainstorms occurring in late autumn and winter. The magnitude and duration of streamflow response to these frontal rainstorms are determined largely by the amount of rain that falls, whether snow is present on the ground, and the amount of antecedent (prior) moisture in the soil. The duration of streamflow originating from high-intensity, short-duration, and mostly convectional rainstorms during the summer and early autumn monsoon season is only a few hours or a few days, depending on the rainfall intensity and amount of the antecedent soil moisture on the watershed.

Sedimentation

Soil particles eroded from hillslopes, entrained in surface runoff, deposited in stream channels, and transported downstream with

Figure 5.1. Perennial streamflow regimes in Arizona are limited in their extent, as shown here in the Beaver Creek watershed in north-central Arizona, approximately 50 miles south of Flagstaff.

flow events are called *sediment*. Therefore, sediment is the product of soil erosion whether it occurs as hillslope erosion, gully erosion, or stream channel erosion. Sediments move continuously into and through the few perennial streams and rivers that remain in Arizona. They are deposited by overland flows into the intermittent and ephemeral stream channels, where they remain until a sufficiently large streamflow-generating event occurs to move these sediment accumulations further downstream (fig. 5.2). Large flushes of sediment movement can occur during large rainfall or snowmelt-runoff events.

Transport of sediments in many of Arizona's streams is cyclic, with alternating periods of aggradation (deposition of sediments) and degradation (removal of sediments) controlling the movement of sediments downstream. The largest sediment loads transported in a single streamflow-generation event are usually associated with the high-intensity, short-duration, and mostly convectional rainfall

Figure 5.2. Sediment accumulation in a stream channel following cessation of ephemeral streamflow in the lower desert of central Arizona.

events occurring at all elevations in late summer through early autumn (Heede et al. 1988; Lopes and Ffolliott 1993; Lopes et al. 2001). Concentrations of sediments in the streamflow originating from snowmelt-runoff events are comparatively low because the energy available to move soil particles in these events is also low. However, much of the annual production of total sediments—including suspended sediments and heavier bedload materials—is attributed to snowmelt-runoff events, largely because these events are the major contributor to annual streamflow amounts.

Suspended sediments represent the largest fraction of the total sediments that move downstream in nearly all of the stream systems in Arizona. As much as 90 percent of the total sediment transported in regimes of high streamflow volumes resulting from flooding events can be suspended sediments (Lopes and Ffolliott 1993; Lopes et al. 2001). Nevertheless, the hydrologic importance of the bedload component of total sediment loads should not be underestimated.

The downstream movement of bedload particles by rolling and sliding in flowing water causes channel erosion and sediment depositions, which can impact the stability of the channel and the nature of the streamflow regimes.

Water Quality

Sediment is a major pollutant of surface water in Arizona. Other physical characteristics of surface water that influence its quality are temperature (thermal) pollution and the level of dissolved oxygen in the water. Chemical pollutants of potential importance include nitrate-nitrogen, phosphate, occasionally sulfate, and heavy metals such as zinc, iron, copper, and manganese. Some chemical constituents contained in surface water are also found in the precipitation that generates streamflow while other constituents are picked up in overland flows of water. Two of the more important microbiological pollutants in surface water are fecal coliform and fecal streptococcus bacteria. Fecal coliform bacteria originates from humans, livestock, and other mammals, whereas the origin of most fecal streptococcus bacteria is from nonhuman mammals. The acceptability of surface water for human consumption, agriculture, or sustaining aquatic habitats is determined largely by the concentrations of these constituents and other water-quality parameters.

6
Plant Communities and Associations

PETER F. FFOLLIOTT and GERALD J. GOTTFRIED

A main feature of Arizona's natural environments are the plant communities and associations in these environments. Picturesque deserts—what many people believe is all that is found in the state—are the dominating landscapes throughout most of the lower elevations. Nearly half of Arizona's land mass is deserts. However, a variety of other plant communities and associations also occur in the state. Diverse grasslands, covering about one-quarter of the state, interface with the deserts and other plant communities and associations at varying elevations throughout the rest of the state. Deep-rooted chaparral (sclerophyllous) shrubs and mostly evergreen (oak) and coniferous (pinyon-juniper) woodlands interface with other communities and associations at intermediate elevations. Extensive stands of mostly coniferous forests are found at higher elevations— including ponderosa pine (*Pinus ponderosa*), mixed conifer, and spruce-fir (*Picea-Abies*) forests in ascending order—with alpine vegetation on the highest mountains. The plant community or association in a particular area largely reflects the combined effects of the climate, geology, and soils of that area.

General descriptions of the plant communities and associations, including their geographic locations, physiographic and climatic characteristics, and commonly observed plant species, are presented in this chapter. The listings of plant species are not intended to be all inclusive. Instead, they illustrate the variety of species commonly encountered within the respective communities and associations. References containing other listings of plant species are included with these descriptions. Common and scientific names of the plants mentioned in the chapter follow those specified by the U.S. Department of Agriculture's Natural Resources Conservation Service (2005). Other taxonomic listings, including Brown (1994), Epple (1995), and Stubbendieck et al. (2003), were consulted when the species in question was not included in this database.

Deserts

The Sonoran, Chihuahuan, Mojave, and Great Basin deserts, covering about 45 percent of the state, are major contributors to Arizona's picturesque landscapes. The Sonoran, Chihuahuan, and Mojave deserts are known as warm-temperature deserts because of their warm and often hot temperatures. In contrast, the Great Basin Desert experiences colder winters and therefore is considered a cold-temperature desert, which often occurs within the rainshadows of mountains. However, temperature is not the sole factor in determining the types of vegetation in these deserts. Precipitation is often a more important factor determining the character of the inhabiting vegetation. This vegetation has gradually evolved through a series of grassland stages to the present-day desertscrub conditions largely in response to the increasing aridity experienced since the Miocene (Brown 1994). As a consequence, the vegetation of the Sonoran, Chihuahuan, and Mojave deserts is often referred to as southwestern desertscrub, while that of the Great Basin Desert is Great Basin desertscrub.

In addition to the plants listed below, more complete listings of the common plants encountered in the deserts of Arizona are found in Brown (1994), Epple (1995), McClaran and Van Devender (1995), Phillips and Comus (2000), McClaran et al. (2003), and Stubbendieck et al. (2003).

The Sonoran Desert

The Sonoran Desert covers much of southwestern Arizona; parts of southeastern California; much of Sonora, Mexico; most of the Baja California peninsula in Mexico; and many of the islands in the Gulf of California. It occupies lower mountain slopes, intervening valleys, and rocky hills. The Sonoran Desert differs from the other North American deserts in that it rarely experiences severe or prolonged frosts, resulting in a lush biota in comparison to the other desert communities.

The Sonoran Desert is the hottest of Arizona's deserts, with temperatures frequently exceeding 100°F in the summer months. Winter freezing occasionally happens. Higher elevations approach 3000 feet, where precipitation can amount to 12 inches annually. At the lowest elevations of about 100 feet in the Yuma Valley along the Lower Colorado River, annual precipitation can be as little as 4 to 5 inches. Precipitation in the eastern part of the Sonoran Desert is biseasonal, with about half falling during the summer monsoon season and the remainder falling in late autumn and throughout the winter. Intermittent water flow in stream systems originating within the Sonoran Desert can result from frontal rainfall events in the winter and, more rarely, from high-intensity convectional rainfall events in the summer.

Lowe (1964a) recognized two principal plant associations in the Sonoran Desert, the paloverde-saguaro (*Parkinsonia-Carnegia*) and the creosote bush–burr ragweed (*Larrea-Ambrosia*) associations. The paloverde-saguaro association is comprised of small-leaved trees and a variety of shrubs, grasses, and cacti (fig. 6.1). This association attains its best development on rocky hills, bajadas, and other coarse-soiled slopes in the uplands between Ajo and Tucson and is characterized by yellow paloverde (*Parkinsonia microphylla*); saguaro (*Carnegia gigantea*); desert ironwood (*Olneya tesota*); crucifixion thorn (either *Canotia holacantha* or *Castela emoryi*); the treelike chollas, particularly jumping cholla (*Opuntia fulgida*); and senita cactus (*Pachycereus schottii*). The distinctive columnar organpipe cactus (*Stenocereus thurberi*) is restricted to southwestern Arizona.

Among the trees in dry arroyos are blue paloverde (*Parkinsonia florida*), velvet mesquite (*Prosopis velutina*), catclaw acacia (*Acacia greggii*), desert willow (*Chilopsis linearis*), and netleaf hackberry (*Celtis laevigata* var. *reticulata*). Smoketree (*Psorothamnus spinosus*) is found in sandy washes. These trees and associated shrub species

Figure 6.1. Small-leaved trees, numerous cacti, and a variety of shrubs, grasses and other herbaceous plants constitute the paloverde-saguaro (*Parkinsonia-Carnegia*) association of the Sonoran Desert in southwestern Arizona.

often form well-developed riparian associations in both the palo-verde-saguaro and creosote bush–burr ragweed associations. Palo-verde, saguaro, and ironwood grow in riparian associations in the Yuma area of southwestern Arizona. Cottonwoods (*Populus* spp.) and willows (*Salix* spp.), which are more typical of riparian trees elsewhere in the state, rarely occur. Shrubs inhabiting a site can be a single species, such as triangle burr ragweed (*Ambrosia deltoidea*) or brittlebush (*Encelia farinosa*), although they are more often clumped in a mixture of 15 or more species in the form of three-, four-, or five-layered understories.

The other principal plant association in the Sonoran Desert, the much simpler creosote bush–burr ragweed association, is comprised mostly of shrub species (fig. 6.2). Trees are mostly absent except for those species inhabiting dry arroyos and washes. Dominant shrubs over extensive areas are creosote bush (*Larrea tridentata*

Figure 6.2. Shrubs are the dominant plants in the creosote bush–burr ragweed (*Larrea-Ambrosia*) association of the Sonoran Desert in southwestern Arizona.

var. *tridentata*), triangle burr ragweed (*Ambrosia deltoidea*), and burrobush (*A. dumosa*) growing either together or alone. Compared to the paloverde-saguaro association, the creosote bush–burr ragweed association occupies less rocky sites, such as mesas, shelving plains, and valleys with gentler slopes. This plant association is interspersed with the paloverde-saguaro association across much of Arizona west of Tucson and Phoenix.

Other plant associations are also scattered throughout the Sonoran Desert landscape. Cattle saltbush (*Atriplex polycarpa*) frequently forms extensive stands across valley bottoms that often have fine-textured, mostly alkaline soils. Isolated remnants of the large forest-like stands of mesquite trees (*Prosopis* spp.), known as mesquite bosques, that grew along major drainage systems such as the Gila River and its tributaries in east-central Arizona are still present. Jojoba (*Simmondsia chinensis*) is found on rocky upland sites in both the paloverde-saguaro and creosote bush–burr ragweed associations. Ocotillo (*Fouquieria splendens*) occurs on elevated sites with rocky shallow soils in both plant associations. These plant species occur in small, nearly pure stands and occasionally in larger stands including other species.

Grasses are not always as noticeable as trees, shrubs, and cacti, but they are relatively numerous in terms of their representative species (Brown 1994; McClaran and Van Devender 1995). Perennial grasses include low turf-grasses, such as curly-mesquite (*Hilaria belangeri*), to the nearly 3-foot-tall Arizona cottontop (*Digitaria californica*) and tanglehead (*Heteropogon contortus*). Other perennial grasses include big galleta (*Pleuraphis rigida*) and bush muhly (*Muhlenbergia porteri*). Among annuals that emerge in late winter and spring and are abundant following good winter rains are Bigelow's bluegrass (*Poa bigelovii*), little barley (*Hordeum pusillum*), and sixweeks fescue (*Vulpia octoflora*). Summer annuals that germinate following the monsoon rains include needle and sixweeks grama (*Bouteloua aristidoides* and *B. barbata*) and littleseed muhly (*Muhlenbergia microsperma*).

Lehmann lovegrass (*Eragrostis lehmanniana*), an aggressive species introduced into the southwestern United States from South Africa in 1913, is a dominant perennial grass on many sites throughout the Sonoran Desert and the region. The noxious and highly invasive buffelgrass (*Pennisetum ciliare*), another introduced African grass, displaces native species throughout many of southwestern desert communities. Red brome (*Bromus rubens*), an annual introduced originally in California in 1848, is also commonly encountered throughout the Sonoran Desert and associated regions.

Among the other herbaceous plants often found in the Sonoran Desert are Gordon's bladderpod (*Lesquerella gordonii*), lyreleaf jewelflower (*Streptanthus carinatus*), Parry's beardtongue (*Penstemon parryi*), sacred thorn-apple (*Datura wrightii*), and several species of desert-thorn (*Lycium* spp.) and buckwheat (*Eriogonum* spp.).

The most diverse cacti flora on Arizona's landscapes grows in the Sonoran Desert (fig. 6.3). Many species representing all of the southwestern cactus forms are present, including species of columnar and

Figure 6.3. Columnar cacti, such as the saguaro (*Carnegia gigantea*) shown here, are among the cacti flora of the Sonoran Desert.

giant cactus (*Carnegia* spp.), barrel cactus (*Ferocactus* spp.), cholla and pricklypear (*Opuntia* spp.), hedgehog and rainbow (*Echinocereus* spp.), and pincushion (*Chaenactis* spp.). The order of decreasing richness of cacti in the flora of Arizona's deserts is the Sonoran, Mojave, Chihuahuan, and Great Basin.

The Chihuahuan Desert

The Chihuahuan Desert is represented in Arizona by a relatively small and isolated area above 3500 feet in the southeastern corner of the state, primarily in Cochise County. It also includes small areas of southern New Mexico and sites along the Rio Grande River in Texas, where it is generally found at lower elevations. Almost 90 percent of the Chihuahuan Desert lies in northern Mexico, including the states of Chihuahua and Coahuila and parts of Durango, Zacatecas, Nuevo León, and San Luis Potosi.

Climatic conditions of the Chihuahuan Desert are intermediate between the desert and grassland communities, and a small change in the precipitation-evaporation relationship can significantly change the vegetation at a locality (Brown 1994). Although the Chihuahuan Desert is the southernmost of the four deserts in Arizona, it lies generally at higher elevations than the other warm deserts of the state, and winter freezes occur as a result. However, the plant community is rich in species despite the low winter temperatures. Summer temperatures can reach 100°F. Annual precipitation ranges from less than 8 inches to more than 10 inches, with about two-thirds falling during the summer monsoon season. Most of the intermittent streamflow in the Chihuahuan Desert originates from frontal rainfall events in the winter. High-intensity convectional rainstorms infrequently produce short-duration streamflow in the summer.

The Chihuahuan Desert is a shrub desert with a physiognomy that is somewhat similar to that of the Great Basin Desert, as shown in figure 6.4. Parts of the San Simon, Sulphur, and San Pedro Valleys are dominated by American tarwort (*Flourensia cernua*), creosote bush, Rio Grande saddlebush (*Mortonia scabrella*), or whitethorn and viscid acacia (*Acacia constricta* and *A. neovernicosa*). These

Figure 6.4. The physiognomy of the Chihuahuan Desert in the southeastern corner of Arizona is somewhat similar to that of the Great Basin Desert in the northwestern part of the state.

shrub species are often intermixed with crown-of-thorns (*Koeberlinia spinosa*), littleleaf sumac (*Rhus microphylla*), Wislizenus' senna (*Senna wislizeni*), ocotillo, and velvet mesquite in its shrubby form. With the exception of creosote bush, ocotillo, and mesquite, these species are among a small group of plants entering southeastern Arizona at the northwestern limits of their respective ranges, with wider distributions elsewhere in the Chihuahuan Desert.

Grasses are generally more abundant in the Chihuahuan than in the Sonoran Desert. Intermingling black, blue, and sideoats grama (*Bouteloua eriopoda*, *B. gracilis*, and *B. curtipendula*) interspersed with stands of curly mesquite and species of threeawn (*Aristida* spp.) are commonly encountered on upland sites. Floodplains or flats supporting stands of tobosagrass (*Pleuraphis mutica*), big sacaton (*Sporobolus wrightii*), and alkali sacaton (*S. airoides*) occur along many drainageways.

Other common herbaceous plants include resinbush (*Viguiera stenoloba*), fiveneedle and Hartweg's pricklyleaf and pricklyleaf dogweed (*Thymophylla pentachaeta, T. pentachaeta* var. *hartwegii,* and *T. acerosa*), sandmat (*Chamaesyce* spp.), desert and Rocky Mountain zinnia (*Zinnia acerosa* and *Z. grandiflora*), and species of sunflower (*Helianthus* spp.) and crinklemat (*Tiquilia* spp.).

The Chihuahuan Desert has a limited representation of cacti and other succulents in comparison to the Sonoran and Mojave Deserts. However, it is richer in this respect than the more northerly Great Basin Desert. Cholla, pricklypear, barrel, and pincushion cacti are represented by the more common species, including walkingstick and Christmas cactus (*Cylindropuntia spinosior* and *C. leptocaulis*), purple pricklypear and cactus apple (*Opuntia macrocentra* and *O. engelmanni*), and candy barrel cactus (*Ferocactus wislizeni*).

The Mojave Desert

The Mojave Desert is a transitional plant community situated between the lower elevation and hotter Sonoran Desert to the south and the higher elevation and cooler Great Basin Desert to the north. It is found in the northwestern parts of Arizona northward from a line between Needles, California, and Congress Junction. The Mojave Desert is more extensively distributed in southwestern California, southern Nevada, and southwestern Utah. It occupies flat to sloping plateaus, plains, and lower mountain slopes interrupted by peaks and moderately incised canyons. Elevations of the Mojave Desert in Arizona are 3500 feet and above.

Summer temperatures in the Mojave Desert approach 100°F. Although freezes occur, they are not as severe as those encountered in the higher Great Basin Desert. Annual precipitation averages about 8 inches, with a winter dominance. Winter precipitation in the form of frontal rainfall events is the expected norm in the Mojave Desert and western part of the Sonoran Desert, and these rainstorms can generate intermittent streamflow.

The most prominent plant in the Mojave Desert is the Joshua tree (*Yucca brevifolia*), which is found in extensive stands above eleva-

Figure 6.5. One of the prominent plants of the Mojave Desert in Arizona is the Joshua tree (*Yucca brevifolia*), which forms extensive stands at the higher elevations of the plant community.

tions of 3000 feet (fig. 6.5). Sonoran Desert species such as paloverde, saguaro, creosote bush, burrobush, ocotillo, and crucifixion thorn are encountered in the southern part of the desert. Great Basin species such as blackbrush occur in the northern limits of its distribution in Arizona. Species not normally considered endemic to the Mojave Desert, such as the devil cholla (*Grusonia emoryi*), are scattered throughout.

Mojave yucca (*Yucca schidigera*) is also conspicuous on many sites, where the species is commonly associated with creosote bush, white bursage, bladdersage (*Salazaria* spp.), brittlebush, chollas, and California barrel cactus (*Ferocactus cylindraceus*). Mojave yucca is generally found at lower elevations than Joshua tree, although the two species frequently occur together on the upper parts of outwash slopes. Banana yucca (*Y. baccata*) grows with Mojave yucca and Joshua tree on many sites in Mohave County.

Other principal plants are creosote bush—the primary shrub

species of the plant community—blackbrush (*Coleogyne ramosissima*), and species of saltbush (*Atriplex* spp.). Mexican bladdersage (*Salazaria mexicana*) is widespread and forms nearly pure stands in some areas. Three species of small trees or shrubs that are mostly restricted to washes in the southern part of the community are catclaw acacia (*Acacia greggii*), desert willow (*Chilopsis linearis*), and western honey mesquite (*Prosopis glandulosa* var. *torreyana*). Other minor plant species are white bursage, bastardsage (*Eriogonum wrightii*), and other shrubby plants such as the commonly encountered broom snakeweed (*Gutierrezia sarothrae*).

Grasses and other herbaceous plants are sparse, and those species present such as big galleta (*Pleuraphis rigida*) and tobosagrass (*P. mutica*) are more commonly associated with other warm desert communities. A variety of ephemeral plants are found in the Mojave Desert, many of which are endemic.

Several forms of cacti are found in the Mojave Desert with one or more of the yucca species, on sites largely restricted to the coarse soils on the gentler outwash slopes. Other cacti are devil cholla (*Grusonia emoryi*); buckhorn, golden, and branched pencil chollas (*Cylindropuntia acanthocarpa, C. echinocarpa,* and *C. ramosissima*); golden, tulip, and beavertail pricklypear (*Opuntia erinacea, O. laevis,* and *O. basilaris*), Engelmann's hedgehog cactus (*Echinocereus engelmannii*); barrel cactus; and desert spinystar (*Mammillaria vivipara* var. *deserti*). Saguaro reaches its northernmost limit in the southern part of the Mojave Desert west of the Hualapai Mountains in southern Mohave County.

The Great Basin Desert

The southeastern limit of the Great Basin Desert lies in the northern part of Arizona, principally west, north, and east of Flagstaff. It also occurs in the extreme northwest of the state near Utah and is well represented in the Arizona Strip north of the Grand Canyon. The northeastern region of the state between the Little Colorado River and the Hopi Reservation, named the Painted Desert by early geologists (Dellenbaugh 1932), is considered a minor subdivision of the larger Great Basin Desert. The widespread exposure of brilliantly colored soil layers on the eroding hills of this region gives this area its name. The Great Basin Desert is the highest of the deserts in the southwestern United States, with much of the community above 6000 feet. It is situated on relatively flat to gently sloping terrain with an interspersion of isolated peaks and incised canyons.

The Great Basin Desert's affinities with cold-temperature plant forms sets it apart from the three warm deserts in Arizona, which have almost exclusive ties with warm-temperature and often subtropical plants. Most of the Great Basin Desert receives less than 10 inches of precipitation annually, with a winter dominance on the westward side of the desert, shifting eastward to a more summer dominance. Streamflow originating from these precipitation events is infrequent.

Trees are almost totally absent in the Great Basin Desert. Rather, intermingling shrub and grass associations that are mostly low in stature, often uniform in density, and consisting of a few species characterize the landscapes of uniform relief (fig. 6.6). The dominant shrub species is big sagebrush (*Artemisia tridentata*). Blackbrush (*Coleogyne ramosissima*), shadscale saltbush (*Atriplex confertifolia*), Mormon tea (*Ephedra viridis*), and greasewood (*Sarcobatus vermiculatus*) are also commonly found in the plant community. Each of these species is found in more-or-less pure stands with only a few grass species present. Other prominent shrubs occasionally encountered are fourwing saltbush (*Atriplex canescens*), black and sand sagebrush (*Artemisia nova* and *A. filifolia*), rubber rabbitbrush (*Ericameria nauseosa*), broom snakeweed, narrowleaf yucca (*Yucca angustissima*), pale desert-thorn (*Lycium pallidum*), stretchberry (*Forestiera pubescens*), and Utah serviceberry (*Amelanchier utahensis*).

The primary grass species in the Great Basin Desert is big sacaton (*Sporobolus wrightii*). Species such as Indian ricegrass and desert needlegrass (*Achnatherum hymenoides* and *A. speciosum*), squirreltail (*Elymus elymoides*), and several species of grama are often interspersed. Introduced species including red brome and cheatgrass (*Bromus tectorum*) are also present.

Other herbaceous plants are often introduced species, such as the

Figure 6.6. Shrub and grass communities of low stature and uniform density dominate the landscape of the Great Basin Desert in northern Arizona.

barbwire Russian thistle and prickly Russian thistle (*Salsola paulsenii* and *S. tragus*), redstem stork's bill (*Erodium cicutarium*), and tall tumblemustard (*Sisymbrium altissimum*).

A few species of cacti grow in the Arizona portion of the Great Basin Desert. Hairspine, grizzlybear, plains, and brittle pricklypear (*Opuntia polyacantha* var. *polyacantha, O. polyacantha* var. *hystricina, O. polyacantha,* and *O. fragilis*) and Whipple cholla (*O. whipplei*) are the most abundant.

Grasslands

Desert, plains, and mountain grassland communities are found on almost 25 percent of Arizona. Desert grasslands attain their best development in the southeastern part of the state; plains grasslands are concentrated in the northeast and scattered elsewhere in the north; and mountain grasslands are natural openings within the coniferous forests of the state. Climatic characteristics vary among the three grassland communities, with wide fluctuations in temperatures and precipitation within each of the communities. Desert grasslands are warm-temperature grasslands, while the plains and mountain grasslands are considered cold-temperature grasslands.

In addition to the plants listed below, more complete listings of plants in the grassland communities of the state were presented by Patton and Judd (1970), Brown (1994), Epple (1995), Stubbendieck et al. (2003), and Fletcher and Robbie (2004).

Desert Grasslands

Desert grasslands are a transitional plant community between the lower elevations of the desert communities and higher elevations of the interior chaparral and woodland communities (Brown 1994; McClaran and Van Devender 1995; Loftin et al. 2000; Robbie 2004). They are largely found in southeastern Arizona but also occur in the northwestern quarter of the state in Mohave County. Desert grasslands extend into the southern half of New Mexico and southward onto the Mexican Plateau to northeast of Mexico City and into Puebla, covering large areas of Durango, Zacatecas, Jalisco, Aguascalientes, and Nuevo León. The lower elevational limit of the plant community is about 3000 feet, and it attains its best development between 4000 and 5000 feet.

Climate of the desert grasslands is generally characterized by warm to hot summers and mild winters. Temperature regimes are similar to those in the adjacent desert communities. Annual precipitation ranges from 10 to 15 inches and occasionally more, falling mostly in a biseasonal (summer and winter) pattern. Streamflow is uncommon, although intermittent events can result from frontal winter rains. Summer streamflow from convectional storms is rare.

The two main plant associations inhabiting the desert grasslands were described by Lowe (1964a): one represented by soaptree yucca (*Yucca elata*) and perennial grasses and the other dominantly tobosagrass landscapes. The two associations are largely found between the grass-covered landscapes and sites supporting varying combinations of shrubs and grasses and—more particularly—on

Figure 6.7. A variety of bunchgrasses and other species of grass cover large areas of the desert grasslands in the southeastern corner of Arizona.

undulating terrain or flat valley bottomlands and swales. Differences in the plant species of the two associations are largely the result of differences in topography, soil characteristics, and the overland water flows that prevail within the similar macroclimate regimes.

Shrub and small-tree forms of mesquite, acacia, pricklypear, cholla, ocotillo, yucca, and agave (*Agave* spp.) are scattered on rocky hills and sloping sites. Common sotol (*Dasylirion wheeleri*) and sacahuista (*Nolina microcarpa*) are shrub forms that dominate local sites with shallow soils. Only the tougher grasses of usually lesser forage value, such as ring muhly (*Muhlenbergia torreyi*), Fendler threeawn (*Aristida purpurea* var. *longiseta*), and low woollygrass (*Dasyochloa pulchella*), are occasionally abundant.

Bunchgrasses, such as black, blue, sideoats, slender (*Bouteloua repens*), and hairy grama, cover extensive landscapes on sites where rocks, shrubs, and cacti are few, the soil is deep, and the grasses are well protected from the impacts of erosion (fig. 6.7). Other intermingling perennial grasses include plains lovegrass (*Eragrostis intermedia*), plains bristlegrass (*Setaria vulpiseta*), sand dropseed (*Sporobolus cryptandrus*), Arizona cottontop, and several species of threeawn. Introduced and invasive Lehmann lovegrass and buffelgrass are dominant on many sites.

Other herbaceous plants include woolly cinquefoil (*Potentilla hippiana*), American vetch (*Vicia americana*), common dandelion (*Taraxacum officinale*), and species of hawkweed (*Hieracium* spp.), pea (*Lathyrus* spp.), and clover (*Trifolium* spp.).

Plains Grasslands

Plains grasslands form a nearly uninterrupted plant cover, and most are found in the eastern half of the state (Brown 1994; Engle and Bidwell 2000; Robbie 2004). This plant community is mostly situated between elevations of 5000 and 7000 feet and intermingled with pinyon-juniper (*Pinus-Juniperus*) woodlands and big, black, and sand sagebrush associations, as shown in figure 6.8. Plains grasslands can also extend upward into the lower elevations of ponderosa pine forests. San Rafael Valley in Santa Cruz County, parts of Sulfur Springs Valley in Cochise County, and Chino Valley in Yavapai County are lower elevation sites of scattered plains grasslands in southern Arizona. Plains grasslands are also found in Utah, Colorado, and New Mexico and extend throughout northeastern Sonora and northwestern Chihuahua in Mexico.

Summer temperatures often reach 95°F in plains grasslands, while freezing temperatures can occur in the winter. Annual precipitation averages about 15 inches, with extremes of 10 and 20 inches. Intermittent streamflow is occasionally generated from frontal winter rainfall events or, less commonly, convectional rainstorms in the summer.

These once-extensive grassland communities have been reduced in extent in northeastern Arizona as a consequence of the invasion of shrub species, particularly in Navajo and Apache Counties. The remaining native grasses include grama, fescue (*Festuca* spp.), dropseed (*Sporobolus* spp.), and muhly (*Muhlenbergia* spp.). James' gal-

Figure 6.8. Plains grasslands intermingle with pinyon-juniper (*Pinus-Juniperus*) woodlands and sagebrush (*Artemisia* spp.) associations in the eastern part of the state.

leta (*Pleuraphis jamesii*) is also a characteristic species whose counterpart in the desert grasslands is tobosagrass. Plains grasslands in southern Arizona often include species of needlegrass (*Achnatherum* spp.). Scattered stands of introduced wheatgrass (*Agropyron* spp.) and brome also occur in these communities.

Among the herbaceous plants other than grasses are woolly plantain (*Plantago patagonica*), Fendler's springparsley (*Cymopterus acaulis* var. *fendleri*), tall tumblemustard (*Sisymbrium altissimum*), and common Russian thistle (*Salsola kali*).

Mountain Grasslands

Mountain grasslands (also known as mountain meadows) are natural openings interspersed throughout ponderosa pine, mixed conifer, and spruce-fir forests. The best development of this plant community is found in the White Mountains in eastern Arizona and

Figure 6.9. Trees and shrubs are not common in mountain grasslands, although isolated individuals occur as a result of seed dispersal from the adjacent forest communities.

throughout the Kaibab Plateau (Fletcher and Robbie 2004). It is also represented on some of the higher mountain ranges in southern Arizona, such as the Pinaleños and Chiricahuas. Mountain grasslands are small in aggregate area. Their interface with forest communities produces an edge effect, where a disproportionately large diversity of plants occurs.

Temperature patterns in mountain grasslands coincide somewhat with those in the adjacent forest communities. The growing season is relatively short, with a frost-free period of only about 90 days. Annual precipitation ranges from about 30 to 35 inches. Streamflow is occasionally generated from snowmelt-runoff events in the late spring, frontal winter rainfalls in the winter, or convectional rainfall in the summer.

Trees and shrubs are not usually found in mountain grassland communities. However, isolated ponderosa pine, Rocky Mountain Douglas-fir (*Pseudotsuga menziesi* var. *glauca*), or Engelmann spruce

(*Picea engelmannii*) can occur on their perimeters as a result of seed dispersals from adjacent forests (fig. 6.9). Drier sites at lower elevations are dominated mostly by grasses and grasslike plants, while wetter sites at higher elevations are more often dominated by large numbers of mostly low-growing forbs.

Characteristic grasses and grasslike plants are alpine timothy (*Phleum alpinum*), pine dropseed (*Blepharoneuron tricholepis*), black dropseed (*Sporobolus interruptus*), mountain muhly (*Muhlenbergia montana*), mountain brome (*Bromus marginatus*), muttongrass and Kentucky bluegrass (*Poa fendleriana* and *P. pratensis*), and species of needlegrass (*Achnatherum* spp.), fescue (*Festuca* spp.), sedge (*Carex* spp.), and rush (*Juncus* spp.).

Other herbaceous plants often give rise to fields of flowers above the carpet of vegetation in the summer. Included in this scenic flora are American vetch, common dandelion, and species of bluebell (*Mertensia* spp.), geranium (*Geranium* spp.), and clover.

Interior Chaparral

Plant communities of largely chaparral shrubs are found in the interior southwestern United States; along the coastal areas of southern California and adjacent Baja California Norte, Mexico; and parts of Chihuahua, Coahuila, and Nuevo León in northern Mexico. Interior chaparral communities of Arizona occur as a discontinuous belt covering 5 to 10 percent of the central part of the state. (Varying criteria used by authorities in delineating the boundaries of interior chaparral communities with other plant communities in the state limit a more precise estimate of its extent.) Chaparral plants extend from the foothills below the Mogollon Rim to south of the Gila River and from the eastern border of the state westward to Mohave County, where they are found as far west as the Hualapai Mountains (Carmichael et. al 1978; DeBano 1999). The lower elevational limit of interior chaparral interfaces with desert or grassland communities, while the upper limit borders or intermingles with the pinyon-juniper woodlands or ponderosa pine forests. Interior chaparral also intermingles with ponderosa pine forests on the slopes of isolated

mountain ranges in the southeastern part of the state. Interior chaparral communities are mostly located between 4000 and 6000 feet in
elevation, although stands of chaparral shrubs are also found as low
as 3500 feet and as high as 7000 feet on isolated mountains often
dissected by steep-walled canyons.

The climatic pattern of interior chaparral is characterized by a
cool and wet winter, followed by a warm and dry spring, then a
second wet period in the summer that lasts until another dry period
in the autumn. The warmest temperatures in the summer vary from
85 to 100°F and higher. Winter temperatures are usually above freezing, although freezing occurs on occasion. The largely biseasonal
precipitation ranges from about 15 inches at the lower elevational
limits to more than 25 inches at the higher elevations. Streamflow
events are mostly attributed to either frontal rains in the winter or
convectional rainstorms in the summer. Snowmelt-runoff events are
less common.

The physiognomy of interior chaparral communities is that of
dense stands of deep-rooted and tough-leaved sclerophyllous shrubs
of uniform height between 3 and 7 feet, occasionally broken by a
taller shrub or short tree (fig. 6.10). More than 50 chaparral species
can be present in these plant communities, but fewer than 15 generally account for most of the plant density on a site (Carmichael et al.
1978; Knipe et al. 1979; Brown 1994; Epple 1995; DeBano 1999).
Sonoran scrub oak (*Quercus turbinella*) is the most commonly encountered chaparral shrub, accounting for 90 percent or more of a
stand's species composition on many sites. Sonoran scrub oak is
likely the same species as coastal sage scrub oak in the coastal California chaparral communities (Brown 1994; DeBano 1999).

Among the common associates of Sonoran scrub oak are Pringle
and pointleaf manzanita (*Arctostaphylos pringlei* and *A. pungens*);
sugar, smooth, and skunkbush sumac (*Rhus ovata, R. glabra*, and *R.
trilobata*); hairy and birchleaf mountain mahogany (*Cercocarpus
montanus* var. *paucidentatus* and *C. montanus* var. *glaber*); desert
ceanothus and deerbrush (*Ceanothus greggii* and *C. integerrimus*);
California and obovate buckthorn (*Frangula californica* and *F. betulifolia* ssp. *obovata*); two species of silktassel (*Garrya wrightii* and

Figure 6.10. Dense stands of sclerophyllous shrubs characterize the physiognomy of the interior chaparral communities in central Arizona.

G. flavescens); hollyleaf redberry (*Rhamnus ilicifolia*); Apache plume (*Fallugia paradoxa*); red barberry (*Mahonia haematocarpa*); catclaw mimosa (*Mimosa aculeaticarpa* var. *biuncifera*); narrowleaf yerba santa (*Eriodictyon angustifolium*); and Mexican cliffrose (*Purshia mexicana*).

Sugar sumac, mountain mahogany, buckthorn, and brickelbush are among disjunctive species of the interior chaparral in Arizona that do not occur in the coastal California chaparral communities (Brown 1994; Epple 1995; DeBano 1999). Several shrubs of the chaparral communities in coastal California and Baja California, Mexico, also occur in the canyons of the interior chaparral communities of Arizona, including California flannelbush (*Fremontodendron californicum*) and California false indigo (*Amorpha californica*).

Grasses are sparse in closed stands, although they can become abundant in open stands and following burns. Black, blue, and side-oats grama; plains lovegrass; bristly wolfstail (*Lycurus setosus*); beard-grass (*Bothriochloa* spp.); and bush muhly are among the more common grasses found in interior chaparral communities. Red brome is an associate on some sites. There are only a few grass species found only in these communities, however.

Other herbaceous plants, also tending to be scarce, include blue-dicks (*Dichelostemma capitatum*), toadflax penstemon (*Penstemon linarioides*), bastardsage (*Eriogonum wrightii*), India and black mustard (*Brassica juncea* and *B. nigra*), yellow sweetclover (*Melilotus officinalis*), purslane (*Portulaca* spp.), and sandmat (*Chamaesyce* spp.).

Other common plants of interior chaparral communities in Arizona were listed by Carmichael et al. (1978), Knipe et al. (1979), Brown (1994), Epple (1995), DeBano (1999), and Stubbendieck et al. (2003).

Woodlands

There are three woodland communities in Arizona: the oak and Mexican oak-pine woodlands of southern Arizona and the pinyon-juniper woodlands mostly in the northern part of the state but with scattered stands in the south as well. While these three communities occupy a somewhat similar ecological niche on nearly 20 percent of the state's landscapes, the overstory of the oak woodlands consists of trees that are largely evergreen oak (*Quercus* spp.); the overstory of the Mexican oak-pine woodlands is an intermingling of evergreen oaks and pine species (*Pinus* spp.); and juniper trees (*Juniperus* spp.) dominate the overstory of the pinyon-juniper woodlands, with scattered pinyons in relatively few numbers. Compositions of grasses and other herbaceous plants also differ among these woodland communities.

In addition to the plants listed below, other listings of plants inhabiting the woodlands of Arizona were presented in Brown (1994), Epple (1995), Gottfried et al. (1995a), Gottfried (1992a, 1999), McPherson (1992, 1997), Ffolliott (1999, 2002), McClaran and McPherson (1999), Gottfried and Pieper (2000), and Stubbendieck et al. (2003).

Oak Woodlands

Oak woodlands—also known as encinal woodlands, from the Spanish word *encinal*, meaning "wholly or mostly oak"—are confined to the southeastern quarter of Arizona. Oak woodlands reach their best development on the foothills of higher mountains, such as the Pinals, Pinaleños, Peloncillos, Galiuros, Santa Catalinas, Baboquivaris, Santa Ritas, Huachucas, and Chiricahuas. Large portions of the woodland communities are also found in New Mexico and Texas, and large stands occur in Sonora, Chihuahua, and Durango in northern Mexico (McPherson 1992, 1997; Ffolliott 1999, 2002; McClaran and McPherson 1999). Oak woodlands occur between 4000 and 6500 feet in elevation throughout their range.

Temperatures range from summer highs approaching and often exceeding 90°F to occasional freezing in the winter. Annual precipitation ranges from 10 to 20 inches, depending on elevation, of which half falls during the summer monsoon season. Winter precipitation—mostly rain but occasionally snow—occurs in the late autumn through spring. Limited streamflow events can originate from frontal rainfall events in the winter or convectional rainstorms in the summer.

Open stand structures with a variety of trees, shrubs, grasses, other herbaceous plants, succulents, and cacti are characteristic of the oak woodlands. Many of the stands at lower elevations are comprised primarily of Emory oak (*Quercus emoryi*), the most common tree species occurring in this woodland community (fig. 6.11). There can also be associates of Arizona white, gray, Mexican blue, and silverleaf oak (*Q. arizonica, Q. grisea, Q. oblongifolia*, and *Q. hypoleucoides*), and occasionally netleaf and Toumey oak (*Q. rugosa* and *Q. toumeyi*) in varying species combinations and spatial arrangements. Alligator juniper (*Juniperus deppeana*) is often abundant on local sites, while redberry juniper (*J. coahuilensis*) and Mexican or border pinyon (*Pinus cembroides* or *P. discolor*) occur sporadically.

Small stands of Arizona cypress (*Cupressus arizonica*) are found at the middle to higher elevations of the oak woodlands. These pockets of trees are largely restricted to north-facing slopes and

Figure 6.11. The Emory oak (*Quercus emoryi*) is the common tree species in the lower elevations of the oak woodlands of southeastern Arizona.

canyon bottoms, where soil moisture is relatively high and temperatures are moderate (Brown 1994; Epple 1995). The subspecies *C. arizonica* ssp. *arizonica* is found mostly south of the Gila River in Cochise, Graham, and Greenlee Counties, while the contiguous subspecies (*C. arizonica* ssp. *glabra*) occurs northward between the Gila River and the Mogollon Rim.

Occurrences of shrubs vary from scattered individuals to landscape dominance. Characteristic species are velvetpod mimosa (*Mimosa dysocarpa*), evergreen sumac (*Rhus virens* var. *choriophylla*), red barberry (*Mahonia haematocarpa*), Schott's yucca (*Yucca schottii*), golden-flowered and Parry's agave (*Agave chysantha* and *A. parryi*), and Palmer's century plant (*Agave palmeri*). Fendler's ceanothus (*Ceanothus fendleri*) and New Mexico locust (*Robinia neomexicana*) occur mostly in the higher elevation ponderosa pine forest but reach their lower elevational limit in the oak woodlands. Interior chaparral shrubs are also found in parts of these communities, including skunkbush sumac and several species of buckthorn, manzanita, mountain mahogany, and silktassel. Shrubs such as skunkbush and a native vine, canyon grape (*Vitis arizonica*), commonly grow beneath the canopies of oak trees, while other shrubs such as pointleaf manzanita more generally grow in the open spaces between trees.

Still other shrubs encountered throughout the plant community are alderleaf mountain mahogany (*Cercocarpus montanus*), Mexican cliffrose, and evergreen sumac. Shrubs and succulents found at desert or grassland edges (ecotones) of the oak woodlands include coralbean (*Erythrina flabelliformis*), both the shrub and small-tree form of velvet mesquite, catclaw mimosa, turpentine bush (*Ericameria laricifolia*), common sotol, banana yucca, and soaptree.

Most of the grasses in the oak woodlands are species of the lower elevation grassland communities or higher elevation forest communities. Blue grama is the most common species throughout the plant community. A few other grasses, such as bullgrass (*Muhlenbergia emersleyi*), little bluestem (*Schizachyrium scoparium*), and squirreltail (*Elymus elymoides*) grow in the more southerly woodland habitat. Others, including common wolfstail and plains lovegrass, are largely found in the interior chaparral and upper parts of the desert and grassland communities.

Herbaceous plants other than grasses are buckbrush and several species of verbena (*Verbena* spp.), globemallow (*Sphaeralcea* spp.), lupine (*Lupinus* spp.), sage (*Salvia* spp.), and mariposa lily (*Calochortus* spp.).

Arizona spinystar (*Escobaria vivipara*), little nipple cactus (*Mammillaria heyderi*), and species from the desert and grassland communities such as saguaro and candy barrel cactus are occasional associates along the lower edge of the oak woodland community. Staghorn cholla (*Cylindropuntia versicolor*) is found mainly in southern Arizona around Tucson. Pricklypear cacti (*Opuntia* spp.) are among the other cacti present.

Mexican Oak-Pine Woodlands

Mexican oak-pine woodlands lie between the oak woodlands below and the ponderosa pine forests above. However, these woodlands are not as widely distributed as the true oak woodlands. Mexican oak-pine woodlands are characterized by the presence of two middle-elevation conifers, Chihuahua and Apache pine (*Pinus leiophylla* var. *chihuahuana* and *P. engelmannii*). These pine species, along with alligator juniper and Mexican or border pinyon, intermingle with one or more of the evergreen oaks, principally silverleaf, Arizona white, or Emory oak.

Other plant forms in this community are largely transitional species. Shrubs, grasses, other herbaceous plants, and other plants inhabiting the oak woodlands often range upward into the Mexican oak-pine woodlands and further upward into the ponderosa pine forests in the mountains of the southeastern Arizona.

Pinyon-Juniper Woodlands

Large areas of the southwestern United States are covered by pinyon-juniper woodlands. This plant community is often separated into the southwestern and the Great Basin pinyon-juniper woodlands on the basis of their species compositions and climatic characteristics, with only the former represented in Arizona. Southwestern pinyon-juniper woodlands surround the higher ponderosa pine forests on or near the Mogollon, Coconino, and Kaibab Plateaus, most characteristically between 5500 and 7000 feet in elevation (Gottfried 1992a, 1999; Gottfried et al. 1995b; McPherson 1997; Gottfried and Pieper 2000). They are also found on flat-topped mesas and plateaus in Navajo and Apache Counties between 5800 and 7200 feet and other northern areas of the state including parts of the Arizona Strip north and west of the Grand Canyon. These woodlands intermingle with the interior chaparral and oak woodlands in the central and southern parts of the state, including Pima and Santa Cruz Counties. They occupy level to gently rolling topography throughout the state, as well as steeper sites on the Kaibab Plateau.

In the pinyon-juniper woodlands of Arizona, high temperatures in the summer can reach 85 to 90°F, with occasional freezing temperatures in the winter. Annual precipitation ranges from 12 to 24 inches, with more than half of the precipitation falling in late autumn through the middle of spring as rain, snow, or combinations

Figure 6.12. Pinyon-juniper woodlands of Arizona are comprised of varying combinations of juniper (*Juniperus* spp.) and pinyon (*Pinus* spp.) tree species, with juniper trees generally more abundant than pinyon on most sites.

of both. Most of the streamflow originating in the pinyon-juniper woodlands is a result of either frontal winter rainstorms or convectional rainfall events in the summer. Melting of the occasional large accumulations of intermittent snowpacks can also result in streamflow events of relatively short durations.

Pinyon-juniper woodlands consist of varying combinations of pinyon (*Pinus* spp.) and juniper (*Juniperus* spp.) trees (fig. 6.12). Juniper trees are generally more abundant than pinyons and frequently occur in almost pure stands below 6800 feet. Dense stands of juniper woodland and more open stands of juniper savanna are minor plant associations of the pinyon-juniper woodlands below 6500 feet in both northern and southern Arizona.

Utah and one-seed juniper (*Juniperus osteosperma* and *J. monosperma*) are the most widespread members of the genus in Arizona, while twoneedle pinyon (*Pinus edulis*) is the most common of the pinyon species. Singleleaf pinyon (*P. monophylla*) is found locally with Utah juniper in the northwestern corner of the state. Rocky Mountain juniper (*J. scopulorum*) is scattered in northeastern and central Arizona. Alligator juniper occurs in the higher elevations of the pinyon-juniper woodland range in central and southern Arizona, and it is also a component of some ponderosa pine forests. Alligator juniper and Mexican or border pinyon are the tree species commonly growing with evergreen oaks in the mountains of southern Arizona. Redberry juniper (*J. coahuilensis*) is occasionally found in this part of the state as well.

Shrub species in the pinyon-juniper woodlands of the northern and central regions of the state include Mexican cliffrose, big and black sagebrush, Utah serviceberry, rubber rabbitbrush, longflower rabbitbush (*Chrysothamnus depressus*), desert sweet (*Chamaebatiaria millefolium*), Mormon tea, Fremont's mahonia (*Mahonia fremontii*), Apache plume, antelope bitterbrush (*Purshia tridentata*), and banana yucca. Shrub species more characteristic of the intermingling southern oak woodlands are absent or weakly represented in the northern pinyon-juniper woodlands. One or more of the shrub species associated with the Great Basin Desert, such as shadscale saltbush (*Atriplex confertifolia*), blackbrush, and winterfat (*Krascheninnikovia lanata*), occur in the lower elevations of the pinyon-juniper woodlands. Interior chaparral shrubs, including Wright's silktassel, littleleaf and curl-leaf mountain mahogany (*Cercocarpus intricatus* and *C. ledifolius*), Sonoran scrub oak, and the shrub form of Gambel oak (*Quercus gambelii*), are commonly present. Shrub species characteristic of the ponderosa pine forest, such as Fendler's ceanothus, are often found in the upper part of these woodlands.

Dominant grass species are blue, sideoats, and black grama; Arizona fescue (*Festuca arizonica*); Indian ricegrass (*Achnatherum hymenoides*); littleseed ricegrass (*Piptatherum micranthum*); prairie Junegrass (*Koeleria macrantha*); sand dropseed (*Sporobolus cryptandrus*); squirreltail; species of threeawn (*Aristida* spp.) and needlegrass (*Achnatherum* spp.); and introduced species of wheatgrass (*Agropyron* spp.).

Other herbaceous plants include broom snakeweed and species of aster (*Aster* spp.), globemallow (*Sphaeralcea* spp.), buckwheat (*Eriogonum* spp.), beardtongue (*Penstemon* spp.), mariposa lily (*Calochortus* spp.), ragwort (*Packera* spp.), paintbrush (*Castilleja* spp.), and sage (*Salvia* spp.).

Among the cacti in the lower elevations of the woodlands are Whipple cholla (*Cylindropuntia whipplei*), beavertail pricklypear (*Opuntia basilaris*), brittle and plains pricklypear, and scarlet hedgehog cactus (*Echinocereus coccineus* var. *coccineus*). Utah grizzlybear pricklypear cactus (*O. polyacantha* var. *erinacea*) grows at the higher elevations.

Forests

Three forest communities are present in Arizona. Along with the intermingling quaking aspen (*Populus tremuloides*) stands, these forests occupy about 6 percent of the state. In ascending elevations they are the ponderosa pine, mixed conifer, and spruce-fir forests. Ponderosa pine forests occur as stands dominated by ponderosa pine trees or stands with other tree species in varying combinations. Mixed conifer forests are comprised of intermingling trees of different species rather than a single species, such as the case with most of the ponderosa pine forests. Mixed conifer forests are also known as Rocky Mountain Douglas-fir forests because this species is a dominant tree (O'Brien 2002). Spruce-fir forests are situated on or near the summits of the highest mountains in Arizona.

In addition to the plants listed below, other plants commonly encountered in the forests and quaking aspen stands of Arizona were presented by Jones (1974), Thill et al. (1983), Brown (1994), Epple (1995), Baker and Ffolliott (1998), Ffolliott and Gottfried (1999), Chambers and Holthausen (2000), and Stubbendieck et al. (2003).

Ponderosa Pine Forests

Ponderosa pine forests are the most widespread forest community in the western United States, ranging in distribution from the Canadian border to Mexico and from the plains states westward into the

Sierra Nevada Mountains. Their occurrence is extensive on the Colorado Plateau of northern, central, and eastern Arizona and western New Mexico, with minor representations in Nevada. Ponderosa pine forests are also present on some of the isolated mountains of southeastern Arizona and southern New Mexico (Schubert 1974; Brown 1994; Covington and Moore 1994; Ffolliott and Baker 1999). These montane forests are situated between 6000 and 7000 feet in elevation at their lower limit and 8500 to 9000 feet at their higher limit, depending on slope exposure. Pure stands of ponderosa pine trees most commonly occur between 7000 and 8000 feet. Ponderosa pine trees can also be found below 7000 feet on moist sites such as north-facing slopes and along stream channels. The lower limit of the species is the lower elevation of the coniferous forests in Arizona, and this lower ecological limit is largely controlled by plant-available soil moisture.

Temperatures throughout the ponderosa pine forests vary from summer highs of 80°F and higher, with cold air from Canada often causing temperatures to drop below 0°F in the winter. There are 20 to 25 inches of precipitation annually, with about 15 inches the minimum amount necessary to maintain the integrity of a ponderosa pine forest. About 45 percent of the annual precipitation falls during the summer monsoons. Winter snow and rainfall events in the northern, central, and eastern regions of Arizona are a major source of annual streamflow in the state. Other streamflow events are caused by large frontal rainfalls in the winter and, occasionally, convectional rainstorms in the summer.

Ponderosa pine is a main tree species on the higher mountains that rise abruptly from valley floors south of the Salt River, where Arizona pine (*Pinus arizonica*) is the primary species. The Rocky Mountain variety of ponderosa pine (*P. ponderosa* var. *scopulorum*) is the tree inhabiting the northern Kaibab Plateau and Mogollon Mesa and other sites on the extensive Colorado Plateau. Other differences in the ponderosa pine forest south of the Salt River and those on the Colorado Plateau of the northern, central, and eastern regions of the state in addition to *Pinus arizonica* and *P. ponderosa* var. *scopulorum* were noted by Lowe (1964a).

Figure 6.13. Scattered stands of ponderosa pine (*Pinus ponderosa*) forests are found on the mountain tops of the sky islands in southern Arizona.

Southern mountains supporting ponderosa pine forests such as the Pinal, Gila, Pinaleño, Santa Catalina, Santa Rita, Huachuca, and Chiricahua Mountains are isolated ranges with relatively steep topography when compared to the plateaus and mesas supporting ponderosa pine forests to the north (fig. 6.13). These southern mountains are populated with many species of plants that have their principal distribution in Mexico and reach their northern limits below and south of the Mogollon Rim (Brown 1994; Gottfried et al. 1995a; Chambers and Holthausen 2000).

Chihuahua and Apache pine are scattered among ponderosa pine trees at lower elevational limits of the forests; alligator juniper is scattered throughout the lower elevations and more xeric habitats; and Rocky Mountain Douglas-fir and southwestern white pine (*Pinus strobiformis*) are common associates on the upper and more mesic sites. Parklike landscapes with shrub, grass, or combinations of these herbaceous plant forms that are typical in northern Arizona are virtually nonexistent in the south. Emory, silverleaf, and netleaf oak and Arizona madrone (*Arbutus arizonica*) are among the evergreen trees inhabiting sites on the lower elevations of the southern mountains. The tree form of Gambel oak, quaking aspen, bigtooth maple (*Acer grandidentatum*), Arizona alder (*Alnus oblongifolia*), and Texas mulberry (*Morus microphylla*) are associates of ponderosa pine trees in the cooler and mesic habitats at higher elevations.

Fendler's buckbrush is a prominent shrub throughout the southern ponderosa pine forests to about 9000 feet. It forms thickets on sites where the forests are open. Oregon boxleaf (*Paxistima myrsinites*) is an inconspicuous shrub seen in the upper elevations of the forests. Wood's rose (*Rosa woodsii* var. *ultramontana*) is found along streams, and New Mexican locust is common throughout most of the plant community. Mountain snowberry (*Symphoricarpos oreophilus*), oceanspray (*Holodiscus discolor*), rock spirea (*H. dumosus*), and orange gooseberry (*Ribes pinetorum*)—also found in the mixed conifer and spruce-fir forests—enter the upper part of the southern ponderosa pine forests between 8500 and 9000 feet. Shrubs of the interior chaparral and oak woodland communities such as redberry buckthorn (*Rhamnus crocea*), California buckthorn (*Frangula californica*), beechleaf frangula (*F. betulifolia*), skunkbush sumac (*Rhus trilobata*), Wright's silktassel, Schott's yucca, deerbrush, and pointleaf and Pringle manzanita are present in the lower parts of these ponderosa pine forests. Smooth sumac (*R. glabra*) occasionally forms thickets on the lower edge of the forests. Here, the facultative species silverleaf and netleaf oak occur in shrub form, and black cherry (*Prunus serotina*) is present in either the shrub or small-tree form.

Mountain muhly, screwleaf muhly (*Muhlenbergia straminea*), and Arizona fescue are the most widespread grass species in the southern ponderosa pine forests. Other grasses and grasslike species include blue grama, squirreltail, pine dropseed, black dropseed, prairie Junegrass, pinyon ricegrass (*Piptochaetium fimbriatum*), muttongrass, and Kentucky bluegrass, and species of brome and sedge. Many of the grasses and grasslike plants occurring in the southern ponderosa pine forests are the same species that inhabit the north, central, and eastern parts of the state.

Other herbaceous plants common to ponderosa pine forests throughout Arizona include grassleaf pea (*Lathyrus graminifolius*), beardlip penstemon (*Penstemon barbatus*), broomlike ragwort (*Senecio spartioides*), pingue rubberweed (*Hymenoxys richardsonii*), and species of cinequefoil (*Argentina* spp.), goldenrod (*Piatradoria* spp.), paintbrush (*Castilleja* spp.), fleabane (*Erigeron* spp.), deervetch (*Lotus* spp.), milkvetch (*Astragalus* spp.), violet (*Viola* spp.), yarrow (*Achillea* spp.), and lupine. Many plants such as western brackenfern (*Pteridium aquilinum*), Fendler's globemallow (*Sphaeralcea fendleri*), Rocky Mountain iris (*Iris missouriensis*), bastard toadflax (*Comandra umbellata*), false pennyroyal (*Hedeoma* spp.), mountain goldenbanner (*Thermopsis montana*), and species of beebalm (*Monarda* spp.) and mullein (*Verbascum* spp.) also occur in the two ponderosa pine forest areas of the state.

In northern, central, and eastern Arizona, ponderosa pine forests are also distributed over extensive areas of flat to rolling plateau terrain (Schubert 1974; Brown 1994; Ffolliott and Baker 1999; Chambers and Holthausen 2000). Their main concentrations stretch along a 220-mile northwest-to-southeast belt from northwest of Flagstaff to the Arizona–New Mexico border, the longest unbroken ponderosa pine forest in the United States (fig. 6.14). Ponderosa pine forests in this part of the state contain fewer trees of other species in comparison to the forests of southern Arizona. However, they are richer in shrub, grass, and other herbaceous species that often extend onto interspersed parklike landscapes. Species of pinyon and juniper trees extend upward from the pinyon-juniper woodlands to intermingle with ponderosa pine trees on sites below 7500 feet elevation. Rocky Mountain Douglas-fir is occasionally encountered at 7000 feet and above. The tree form of Gambel oak is scattered throughout the ponderosa pine forests. Quaking aspen is found in isolated stands on burned or harvested sites mostly above 7500 feet. New Mexican locust occurs throughout these forests.

Among the characteristic shrub species are buckbrush, desert sweet (*Chamaebatiaria millefolium*), the shrub form of Gambel oak,

Figure 6.14. Ponderosa pine (*Pinus ponderosa*) forests of northern, central, and eastern Arizona are concentrated in a belt from northwest of Flagstaff to the Arizona–New Mexico border.

Utah fendlerbush (*Fendlerella utahensis*), wax currant (*Ribes cereum*), common elderberry (*Sambucus nigra* ssp. *canadensis*), greenleaf manzanita (*Arctostaphylos patula*), Parry's rabbitbrush (*Ericameria parryi*), big and black sagebrush, Mexican cliffrose, Apache plume, and littleleaf mock orange (*Philadelphus microphyllus*). Shrub species inhabiting the mixed conifer and spruce-fir forests at higher elevations are also found at the upper limits of the ponderosa pine forests, including mountain ninebark (*Physocarpus monogynus*), grayleaf red raspberry (*Rubus idaeus* ssp. *strigosus*), common juniper (*Juniperus communis* var. *saxatilis*), oceanspray, creeping barberry (*Mahonia repens*), fivepetal cliffbush (*Jamesia americana*), and Oregon boxleaf. Some shrubs, such as species of *Artemisia*, are locally abundant at lower elevations of the forests, as seen on the South Rim of the Grand Canyon and the Navajo Reservation.

Grasses inhabiting the northern, central, and eastern ponderosa pine forests include blue and sideoats grama, mountain muhly, Arizona fescue, spike muhly, pine dropseed, squirreltail, mountain and smooth brome (*Bromus marginatus* and *B. inermis*), deergrass (*Muhlenbergia rigens*), prairie Junegrass, little bluestem, and several species of threeawn (*Aristida* spp.) and wheatgrass (*Pascopyrum* spp.). Blue grama is common in forest openings.

Other herbaceous plants in addition to those mentioned in the discussion of the southern ponderosa pine forests are Fendler's meadow-rue (*Thalictrum fendleri*), American vetch, purple locoweed (*Oxytropis lamberti*), and white sagebrush (*Artemisia ludoviciana*).

Mixed Conifer Forests

The mixed conifer forests are located on the higher mountain ranges on the Colorado Plateau in the northern, central, and eastern parts of the state (Rich and Thompson 1974; Thill et al. 1983; Gottfried 1992b; Gottfried and Ffolliott 1992; Gottfried et al. 1999; Chambers and Holthausen 2000). One of their best developments is found in the White Mountains of eastern Arizona southwest of Springerville and extending to the Mogollon Rim. Mixed conifer forests are also well developed on some of the higher mountains in southern Arizona. These forests generally top-out between 9,000 and 10,000 feet on the Santa Catalina, Santa Rita, Huachuca, Chiricahua, and Pinaleño Mountains. The existence of Engelmann spruce in the Chiricahua Mountains represents its southernmost occurrence in the United States.

Because they occur at higher elevations, temperature regimes in the mixed conifer forests are cooler than those in the ponderosa pine forests. Annual precipitation averages more than 30 inches, almost equally divided between summer and winter seasons. Seventy-five percent and occasionally more of the annual streamflow from watersheds in the mixed conifer forests comes from winter snowfalls. Streamflow-generating mechanisms in mixed conifer forests are similar to those in ponderosa pine forests.

Rocky Mountain Douglas-fir and white fir (*Abies concolor*) are the principal species on most sites in these forests (Jones 1974; Thill et al. 1983; Gottfried 1992b; Ffolliott and Baker 1999). Rocky Mountain Douglas-fir has the wider tolerance of the two species, allowing it to extend to lower elevations, and predominates on south-facing exposures. White fir tends to dominate on northerly exposures. Engelmann spruce and occasionally blue spruce (*Picea pungens*) intermingle with Rocky Mountain Douglas-fir and white fir trees on some sites. Corkbark fir (*Abies lasiocarpa* var. *arizonica*) can be present with white pines at the higher elevations (fig. 6.15), with limber pine (*Pinus flexilis*) in the northern mixed conifer forests being replaced by southwestern white pine (*Pinus strobiformis*) in the south. Scattered ponderosa pine trees are confined to ridges and southerly exposures except along the lower edge of the forests, where mixed conifer and ponderosa pine forests often merge in a continuum. Mixed conifer forests also intermingle with the higher elevation spruce-fir forests, leading some people to consider the two forest communities together for management purposes, which is generally a mistake because of their differing physiognomy (Moir and Ludwig 1979).

Individual quaking aspen trees and stands (see below) are scattered throughout the mixed conifer forests. The tree form of Gambel oak, boxelder (*Acer negundo*), blue elderberry (*Sambucus nigra* ssp. *cerulea*), and water birch (*Betula occidentalis*) also occur throughout the forests, with Rocky Mountain maple (*Acer glabrum*) at higher elevations and New Mexico locust at lower elevations.

Many of the common species of shrubs, grasses, grasslike species, and other herbaceous plants found in mixed conifer forests are shared with either the lower elevation ponderosa pine forests or the higher elevation spruce-fir forests. Among the shrub species are Fendler's buckbrush, New Mexico locust, red elderberry (*Sambucus racemosa*), and snowberry (*Symphoricarpos* spp.). Representative grasses include Arizona and red fescue (*Festuca arizonica* and *F. rubra*), mountain muhly, and spike trisetum (*Trisetum spicatum*). Sedges occasionally dominate in the understory on some sites. Other herbaceous plants are Oregon boxleaf and several species of gentian (*Gentiana* spp.), primrose (*Primula* spp.), fleabane, and vetch.

Figure 6.15. Engelmann and blue spruce (*Picea engelmannii* and *P. pungens*) intermingle with Rocky Mountain Douglas-fir (*Pseudotsuga menziesi* var. *glauca*) and white fir (*Abies concolor*) in the mixed conifer forests of Arizona.

Quaking Aspen Stands

Quaking aspen trees frequently intermingle with a variety of coniferous tree species as scattered individuals, but the species' occurrence in largely pure stands (fig. 6.16) is the emphasis of this discussion. Quaking aspen stands form on burned, harvested, or otherwise disturbed sites within the mixed conifer forests; the upper part of the ponderosa pine forests; and throughout the higher elevation spruce-fir forests (Jones 1974; Brown 1994). Quaking aspen stands are found in the White Mountains, on the lower slopes of the San Francisco Peaks, throughout the Kaibab Plateau, and across the higher mountain ranges in southern Arizona. These stands can be well developed in their structure prior to the establishment of coniferous tree species that successively follow quaking aspen in time after a disturbance.

A variety of shrubs, grasses and grasslike plants, and other herbaceous plants grow in the understory of quaking aspen stands. Many of these plant species are closely associated with the early successional stages of the conifer forests in which these stands are established, with a greater diversity of species often found on more open sites.

Spruce-Fir Forests

Spruce-fir forests are situated on the summits of the San Francisco Peaks and the Pinaleño and Chihuahua Mountains, within the White Mountains of eastern Arizona, and on the top of the Kaibab Plateau (Jones 1974; Moir and Ludwig 1979; Brown 1994; Chambers and Holthausen 2000). Elevations of these forests are 8500 to 9000 feet at their lower limit to approximately 11,500 feet at their upper limit.

Summer temperatures are cooler than the lower elevation mixed conifer forests, and winter temperatures are colder. Spruce-fir forests receive 25 to 35 inches of precipitation annually, much of it in the form of winter snowfalls. Streamflow is generated by the same mechanisms as those occurring in the other forest communities of the state.

Figure 6.16. Quaking aspen (*Populus tremuloides*) trees become established on disturbed sites in the mixed conifer forest, upper elevations of ponderosa pine (*Pinus ponderosa*) forests, and throughout the higher elevation spruce-fir forests.

Engelmann spruce is the dominant tree throughout most of the spruce-fir forests in Arizona, with corkbark fir a common companion (fig. 6.17). Rocky Mountain Douglas-fir and white fir occur with Engelmann spruce on Baldy Peak and other sites in the White Mountains as well as on the Kaibab Plateau. Blue spruce is a frequent associate of these tree species on the extensive Kaibab Plateau. Quaking aspen occurs as a serial species in small clumps or larger stands on disturbed sites, as it does in the mixed conifer forests. Limber pine is scattered throughout spruce-fir forests. Rocky Mountain maple, bitter cherry (*Prunus emarginata*), Bebb and Scouler's willow (*Salix bebbiana* and *S. scouleriana*), and thinleaf alder (*Alnus incana* ssp. *tenuifolia*) are among the intermingling tree species where the canopies of spruce-fir forests are not completely closed.

Characteristic shrubs include whortleberry (*Vaccinium myrtillus*), creeping barberry (*Mahonia repens*), twinberry honeysuckle (*Lonicera involucrata*), shrubby cinquefoil (*Dasiphora fruticosa* ssp. *floribunda*), common juniper, red elderberry, and a few species of currant. Most of the shrubs disappear from the landscapes of the spruce-fir forests before the tree line is reached. Exceptions are gooseberry currant (*Ribes montigenum*) and dwarf juniper, which occasionally extend beyond the tree line on the San Francisco Peaks.

Many of the grasses present in the lower spruce-fir forests are the same species found in the mixed conifer forests and the higher elevations of ponderosa pine forests. Other herbaceous plants frequently encountered in the lower part of these forests are also common to the mixed conifer forests. Grasses, grasslike plants, and other herbaceous plants growing in the higher spruce-fir forests between 10,500 and 11,500 feet are often species shared with the higher-elevation alpine plant communities. The higher elevation grasses and grasslike plants include alpine timothy, red fescue (*Festuca rubra*), spike trisetum (*Trisetum spicatum*), and several species of sedge and rush. Other herbaceous plants at these elevations are species of columbine (*Aquilegia* spp.), false hellebore (*Veratrum* spp.), sneezeweed (*Helenium* spp.), baneberry (*Actaea* spp.), lousewort (*Pedicularis* spp.), primrose, gentian, and violet.

Figure 6.17. The Engelmann spruce (*Picea engelmannii*) is the dominant tree species in the spruce-fir forests in Arizona, with corkbark fir (*Abies lasiocarpa* var. *arizonica*) and other coniferous species common companions.

An Alpine Plant Community

An isolated alpine plant community is found on the top of the San Francisco Peaks near Flagstaff. These peaks rise to nearly 12,670 feet in elevation, the highest point in Arizona. The alpine plant community is confined to an area above tree line, which occurs on the San Francisco Peaks at approximately 11,480 feet (fig. 6.18). While this limited representation is the only true portion of an alpine plant community in the state, a few alpine plant species are found on Baldy Peak in the eastern White Mountains and on Mount Graham in the Pinaleño Mountains south of the Gila River (Brown 1994; Epple 1995; Chambers and Holthausen 2000). Climatic conditions above tree line on the San Francisco Peaks are characterized by cooler temperatures and higher precipitation amounts compared to the other plant communities in the state.

Stunted and gnarled or prostrate Engelmann spruce and bristlecone pine (*Pinus aristata*) grow singly or scattered in small patches at the tree line on the San Francisco Peaks. Gooseberry currant and common juniper are also found at tree line and sheltered sites immediately above it. Several species are mat-forming cushion plants, and all are dwarf in form. Grasses and grasslike plants, other low-growing herbaceous plants, mosses, and lichens are found with ferns and liverworts. Two species that appear to be endemic to the San Francisco Peaks are San Francisco Peaks ragwort (*Packera franciscana*) and elephanthead lousewort (*Pedicularis groenlandica*).

Three plant associations were recognized by Lowe (1964a) in his discussion of the state's alpine plant community: the alpine tundra rock field, alpine tundra meadow, and talus associations. Southwestern showy sedge (*Carex bella*) is a common sedge in the alpine tundra rock field; spiked woodrush (*Luzula spicata*) is the common rush; and alpine fescue (*Festuca brachyphylla*), timberline bluegrass (*Poa glauca* ssp. *rupicola*), and spike trisetum (*Trisetum spicatum*) the common grasses. Most of the flowering plants found above tree line inhabit the alpine tundra rock field. Crustose and foliose lichens occur on rock surfaces.

The alpine tundra meadow association on the San Francisco Peaks

Figure 6.18. An isolated alpine plant community is found above the tree line on the often snow-capped San Francisco Peaks near Flagstaff.

is dry in comparison to the wet meadows within the alpine tundra found in the higher Colorado Rockies. The *Geum rossii* var. *turbinatum–Carex* complex is prevalent. Other mat-forming plants include creeping sibbaldia (*Sibbaldia procumbens*) and moss campion (*Silene acaulis*). Alpine timothy and nodding bluegrass (*Poa reflexa*) are largely restricted to this plant association. Southwestern showy, blackandwhite, and ebony sedges (*Carex bella, C. albonigra,* and *C. ebenea*); spiked woodrush (*Luzula spicata*); and Drummond's rush (*Juncus drummondii*) are encountered. Several species of mosses and one liverwort (*Lophozia* spp.) are also present.

The talus association is situated on sites with little soil development, freezing and near-freezing winter temperatures, and desiccating winds; therefore, it is too harsh for vascular plants to become established. However, lichens and mosses often abound.

Listings of other plant species in the alpine plant community of the state are found in Brown (1994), Epple (1995), and Stubbendieck et al. (2003).

Riparian Associations

Riparian associations are situated along the banks of perennial, intermittent, and ephemeral streams and rivers within all of the plant communities and associations in Arizona, with the exception of the alpine plant community (Brown 1994; DeBano and Baker 1999; Cartron et al. 2000; Baker et al. 2004). The abruptness of the transition between the terrestrial and aquatic interfaces that define these riparian corridors are site specific regardless of where they occur. While riparian associations occupy less than 2 percent of the total land area in Arizona, they are often the most appreciated of all of the ecosystems in the state in providing opportunities for hiking, picnicking, bird watching, and camping. Because compositions of inhabiting plant species change with elevation, several authorities (Szaro 1989; Baker et al. 1998, 2004; DeBano and Baker 1999) recognize three elevational strata of the riparian associations: those below 3500 feet, between 3500 and 7000 feet, and above 7000 feet in elevation.

Riparian associations below 3500 feet are found on broad alluvial floodplains and often on terraced bottomlands (fig. 6.19). They support relatively high densities of deep-rooted trees including Arizona sycamore (*Platanus wrightii*), Fremont cottonwood (*Populus fremontii*), velvet mesquite, the introduced five-stamen tamarisk (*Tamarix chinensis*), species of paloverde (*Parkinsonia* spp.), and a large variety of grasses, grasslike plants, and other herbaceous plants, including several species of bulrush (*Blysmus* spp.) and spikerush (*Eleocharis* spp.). Only a few of the formerly widespread mesquite bosques remain, and most of the large cottonwood and willow galleries have been replaced by saltcedar (*Tamarix ramosissima*) and Russian olive (*Elaeagnus angustifolia*). Isolated Emory, Arizona white, Mexican blue, and silverleaf oak finger down the riparian corridors onto the deserts as far as 1000 feet below the normal concentrations of oak woodlands, and species of chaparral shrubs are found at elevations of 2000 feet and lower.

Figure 6.19. Low-elevation riparian associations occur along Sonoita Creek in southern Arizona.

Riparian associations between 3500 and 7000 feet contain the greatest diversity of plant species and the most canopy cover of the three riparian associations (fig. 6.20). Sycamore, willow, cottonwood, velvet ash (*Fraxinus velutina*), and Arizona walnut (*Juglans major*) are commonly found in intermingling combinations as narrow stringers along the mainly intermittent or ephemeral streams. Intermingling oak or pinyon-juniper woodlands or interior chaparral communities often occupy the upland slopes surrounding these middle-elevation riparian associations. A diversity of shrubs, grasses and grasslike plants, and other herbaceous plants—many of which reflect the herbaceous understories on the adjacent hillslopes—occur in this elevational strata.

Willow, black cherry, Rocky Mountain maple, boxelder, Arizona alder, and a variety of coniferous species dominate the tree overstories of riparian associations above 7000 feet in elevation. Shrubs, grasses and grasslike plants, and other herbaceous plants represent the most mesic of the species inhabiting plant communities sur-

Figure 6.20. Middle-elevation riparian associations are found along the ephemeral Dry Beaver Creek in north-central Arizona.

rounding the riparian corridors at these elevations. Mountain grasslands occupy streamside sites with excessive plant-available soil moisture in some locations (fig. 6.21). The uplands surrounding these riparian ecosystems are often characterized by ponderosa pine, mixed conifer, or spruce-fir forests, with the occurrence of these plant communities depending largely on the ecological gradients of the area.

Composite listings of the plant species inhabiting the riparian associations of Arizona were presented by Campbell and Green (1968), Brown (1994), Busch (1995), Epple (1995), Cartron et al. (2000), Snyder et al. (2002), Stromberg (2002), McLaughlin (2004), and others.

Wetland Associations

Wetland associations are areas inundated or saturated by surface water or groundwater at a frequency and duration sufficient to sup-

Figure 6.21. Upper-elevation riparian associations occur within the mountain grasslands along the White River in eastern Arizona.

port vegetation adapted to saturated soil conditions. As a consequence, these associations possess vegetation and soils that distinguish them from adjacent uplands. Scattered wetlands of three types are found in Arizona: riverine wetlands, cienegas, and wet meadows (Hendrickson and Minckley 1984; Minckley and Brown 1994).

Riverine wetlands typically occur as oxbows, behind natural levees, and along stream margins at elevations below 3000 feet (fig. 6.22). These wetlands are often transitory in nature and form when groundwater becomes perched above a layer of clay (Mitsch and Gosselink 1993) or where larger sediments that are deposited by floods impound a stream. Flow of water through riverine wetlands is slow, although it tends to be continuous. Trees are scarce and mostly limited to species that tolerate saturated soils, such as Goodding's and narrowleaf willow (*Salix gooddingii* and *S. exigua*). Other common plants include southern cattail (*Typha domingensis*), chairmaker's bulrush (*Schoenoplectus americanus*), giant reed (*Arundo donax*), common reed (*Phragmites australis*), arrowweed (*Pluchea*

Figure 6.22. This riverine wetland association grows along the edge of the Colorado River.

sericea), and other species that thrive in fluctuating water levels. Isolated riverine wetlands are found along some of the larger river systems in the state, such as the Gila and Lower Colorado.

Cienegas are found in the middle elevations from 3500 to 6500 feet. They often occur in headwater situations where groundwater rises to the surface on numerous sites, resulting in a well-watered flat or valley. Cienegas trap organic materials from adjacent ecosystems and are generally sites with permanently saturated and highly organic soils with reduced oxygen levels. Trees are limited to species that can tolerate saturated soils including species found in riparian associations at these elevations. Most of the plants inhabiting cienegas are low, shallow-rooted, and semi-aquatic sedges, rushes, some grasses, and rarely cattails. Stands of sacaton on adjacent flatlands are also encountered. Watercress (*Nasturtium officinale*) and whorled marsh-pennywort (*Hydrocotyle verticillata*) can be locally abundant.

Wet meadows at elevations above 6500 feet are found on hydric soils, experience low-velocity surface and subsurface water flows, and are subject to freezing and thawing cycles. These wetlands often occur as depressions and, therefore, tend to be lentic habitats (still waters) fed by seepage or precipitation, or lotic habitats (with moving water) bordering headwater streams. Wet meadows should not be confused with the drier mountain grasslands also found at similar elevations. While grasses, sedges, and rushes are the dominant plants in the community, some sites also support shrubs such as thinleaf alder (*Alnus incana* ssp. *tenuifolia*), currant, and willows.

Threatened and Endangered Plants

Species listed by federal or state authorities as threatened or endangered are scattered throughout Arizona's plant communities and associations. Endangered plant species are designated as such when the authorities believe that they will become extinct in the near future unless special care is taken in protecting populations from further loss. Threatened species are those likely to become endangered in the foreseeable future if left unprotected. The listings of threatened and endangered plants by federal or state authorities are almost always dated, in that species are continuously being added to or removed from these listings.

Among endangered plants in the listing compiled by Kurpis (2002) are the Arizona century plant (*Agave arizonica*), Kearney's bluestar (*Amsonia kearneyana*), sentry milkvetch (*Astragalus cremnophylax* var. *cremnophylax*), long-tubercle beehive cactus (*Coryphantha robustispina* ssp. *robustispina*), Nichol's echinocactus (*Echinocactus horizonthalonius* var. *nicholii*), Arizona hedgehog cactus (*Echinocereus coccineus* var. *arizonicus*), Schaffner's grasswort (*Lilaeopsis schaffneriana* var. *recurva*), Brady's hedgehog cactus (*Pediocactus bradyi*), Navajo pincushion cactus (*Pediocactus peeblesianus*), and reclusive lady's tresses (*Spiranthes delitescens*). Threatened plant species include Welsh's milkweed (*Asclepias welshii*), Navajo sedge (*Carex specuicola*), Jones' waxydogbane (*Cycladenia humilis* var. *jonesii*), Siler's pincushion cactus (*Pediocactus sileri*), and San Francisco Peaks ragwort (*Packera franciscana*). This plant listing by Kurpis is not necessarily all-inclusive. Nevertheless, the species listed are indicative of the diversity of threatened and endangered species in the state's plant communities and associations.

7
Vascular Flora

STEVEN P. McLAUGHLIN

The diversity, composition, and other floristic relationships of the vascular flora of Arizona are discussed in this chapter. Comparisons are made with those of adjacent areas and across broader geographic scales to understand the biota of an area. This chapter is based on a data set of 245 local floras from Alaska, Canada, the lower 48 states, and Mexico to define floristic elements and areas for North America and analyze other patterns of distribution of vascular plants at the continental scale (McLaughlin 2007). This data set was also used to illustrate phytogeographic patterns of the Arizona flora.

Diversity and Composition

The milestone book *Arizona Flora* (1960), by Thomas H. Kearney and Robert H. Peebles, remains the most comprehensive floristic reference on vascular plants of the state in listing 3370 native and naturalized species, though it was published more than 45 years ago. To update this book where necessary, representatives from the herbarium of Arizona State University and the Arizona-Nevada Academy of Science solicited contributions from authorities on selected families, genera, and species for the book *Vascular Plants of Arizona*, which included these taxonomic groupings. Some of these contributions have been published in the *Journal of the Arizona-Nevada Academy of Science* and are also available online (http://lifesciences .asu.edu/Herbarium/vpap.html). More recently, Leslie R. Landrum of Arizona State University established *Canotia*, an online journal to publish these contributions as well (http://lifesciences.asu.edu/

Table 7.1 Summary of the diversity of Arizona's vascular flora

Group	Families	Genera	Native species	Naturalized species	Total species	Native subspecific taxa
Pteridophytes	15	31	103	0	103	3
Gymnosperms	3	7	29	0	29	1
Dicotyledons	114	857	2723	286	3009	446
Monocotyledons	24	200	597	124	721	52
Total taxa	156	1095	3452	410	3862	502

Herbarium/canotia.html). Some of the largest families, including Asteraceae, Poaceae, Fabaceae, Cyperaceae, Scrophulariaceae, Boraginaceae, Hydrophyllaceae, and Onagraceae, have yet to be published, however.

Two continental-scale floristic projects are other sources for information on Arizona's flora. Kartesz (1999) provided a synonymized checklist of vascular plants for the United States, Canada, and Greenland. This reference lists 3555 native species for Arizona (J. T. Kartesz, personal communication, 1999). Twelve of the planned 30 volumes of the *Flora of North America* have been published online (http://www.fna.org/FNA/index.html).

An estimate of the diversity of Arizona's flora based on Kearney and Peebles (1960) and available information from the *Vascular Plants of Arizona* and the *Flora of North America* is presented in table 7.1. This estimate of 3452 native species is larger than the 3370 total species in Kearney and Peebles (1960) but still less than the 3555 given in Kartesz (1999), with the number of naturalized exotic species difficult to determine. The estimate of 410 for these latter species represents only 10.6 percent of the total of the flora in the state, a comparatively low proportion within the United States.

Gentry (1986) mapped the number of endemic species for the United States by state. His estimate of 164 endemics in Arizona is likely an underestimate, however, because it excludes a number of recently described species. The percentage of Arizona's native species that is endemic (4.6 percent) is higher than that found in most of the states

Table 7.2 Largest families of Arizona's vascular flora ranked
by the number of native species

Family[1]	Native species	Naturalized species	Total species
Asteraceae (Compositae)*	559	46	605
Poaceae (Gramineae)	306	116	422
Fabaceae (Leguminosae)	306	32	338
Scrophulariaceae	121	10	131
Cyperaceae	112	0	112
Brassicaceae (Cruciferae)	92	32	124
Polygonaceae	86	11	97
Euphorbiaceae	84	8	92
Boraginaceae	76	3	79
Cactaceae	76	2	78
Onagraceae	72	0	72
Rosaceae	70	3	73
Polemoniaceae	60	2	62
Lamiaceae (Labiatae)	54	10	64
Hydrophyllaceae	54	3	57
Apiaceae (Umbelliferae)	53	15	68
Malvaceae	53	5	58
Solanaceae	51	7	58
Ranunculaceae	50	3	53

*Family names in parentheses are those that differ from Kearney and Peebles
1960.

but is lower than that of other southwestern states such as California
(27.7 percent), Texas (8.2 percent), and Utah (5.7 percent).

The largest families of vascular plants in the state are ranked by
the number of native species in the families in table 7.2. The family
circumscriptions presented in this table follow those of Cronquist
(1988), which are largely the same as presented in Kearney and
Peebles (1960) and Kartesz (1999), instead of those presented in
the recent phylogenetic summaries of the Angiosperm Phylogeny
Group (2003). The *Vascular Plants of Arizona* and the *Flora of North
America* also follow the system of Cronquist (1988) in their family

circumscriptions. Classifications based on cladistic analyses of molecular (RNA or DNA) data have focused on the family and ordinal levels. The Angiosperm Phylogeny Group (2003) system allows for alternative family circumscriptions. Families such as Araliaceae, Hydrophyllaceae, Hypericaceae, and Valerianaceae might or might not be recognized.

The sunflower family (Asteraceae), grass family (Poaceae), and legume family (Fabaceae) are the three largest families in Arizona, as they are throughout most of North America. Unlike most other states in the country, however, Arizona has a higher proportion of species in those families found in warmer and drier climates, including Polygonaceae, Euphorbiaceae, Boraginaceae, Cactaceae, Onagraceae, Polemoniaceae, Hydrophyllaceae, Malvaceae, and Solanaceae. While not listed in table 7.2, smaller families such as Amaranthaceae, Chenopodiaceae, Crossosomataceae, Ephedraceae, Loasaceae, Nyctaginaceae, Pteridaceae, and Zygophyllaceae are also comparatively well represented.

The 25 largest genera in Arizona's vascular flora, in terms of number of native species, are listed in table 7.3. The "mean range" values presented in this table are the average number of local floras for the species of each genus based on the author's data set of 245 North American local floras (McLaughlin 2007). These mean ranges provide more precise descriptions of range sizes than qualitative terms such as *narrow* or *widespread*. Other entries in the table are the numbers of species in each genus found in five geographic subdivisions of the state (see fig. 7.1 below). The genera shown in table 7.3 represent almost one-quarter of the total native flora of Arizona. Several of these genera were circumscribed differently in Kearney and Peebles (1960), Kartesz (1999), and other floristic references including the *Vascular Plants of Arizona* and the *Flora of North America*.

The abbreviations *s.l.* (*sensu lato*) and *s.s.* (*sensu stricto*) mean "in the broad sense" and "in the narrow sense," respectively. In table 7.3, *Euphorbia s.l.* is circumscribed to include both *Chamaesyce* and *Poinsettia*, which are often elevated to segregate genera. *Erigeron s.s.* excludes species currently recognized as *Conyza, Laennecia*, and *Trimorpha. Dalea s.s.* excludes *Psorothamnus* and *Marina. Brickellia s.l.*

Table 7.3 The 25 largest genera (based on number of native species) of Arizona's vascular flora and five geographic subdivisions of the state

Genus	Family	Native species	Mean range	Geographical subdivision[1] NE	CH	CA	SW	SE
Astragalus	Fabaceae	78	4.1	48	19	17	18	14
Eriogonum	Polygonaceae	53	6.5	34	10	27	18	14
Muhlenbergia	Poaceae	46	12.1	22	17	19	8	32
Carex	Cyperaceae	45	21.8	7	32	9	1	23
Euphorbia s.l.	Euphorbiaceae	43	14.3	16	11	25	19	38
Phacelia	Hydrophyllaceae	43	3.8	28	3	10	19	9
Penstemon	Scrophulariaceae	41	4.6	21	13	13	6	15
Erigeron s.s.	Asteraceae	34	7.9	15	17	12	4	17
Cryptantha	Boraginaceae	33	7.2	19	3	9	16	9
Dalea s.s.	Fabaceae	30	11.9	6	5	8	3	23
Cyperus	Cyperaceae	28	18.8	4	2	10	10	23
Opuntia s.l.	Cactaceae	28	9.9	12	0	10	16	11
Asclepias	Asclepiadaceae	27	13.0	11	7	9	4	18
Brickellia	Asteraceae	24	10.1	6	4	10	7	16
Lupinus	Fabaceae	24	7.1	11	7	11	5	9
Mentzelia	Loasaceae	23	5.4	10	2	6	9	9
Senecio s.l.	Asteraceae	23	7.4	6	12	8	3	11
Juncus	Juncaceae	23	36.5	11	12	11	10	14
Atriplex	Chenopodiaceae	20	7.3	10	1	3	10	6
Camissonia	Onagraceae	20	4.3	10	0	6	11	3
Gilia s.s.	Polemoniaceae	19	5.6	9	1	7	11	4
Oenothera s.s.	Onagraceae	19	14.3	11	7	13	4	13
Ranunculus	Ranunculaceae	19	24.6	6	16	6	1	4
Potentilla	Rosaceae	19	22.3	3	16	2	0	8
Salix	Salicaceae	18	31.1	11	13	11	4	7
	Totals	780		347	230	272	217	350

Notes: Mean range values represent the average number of North American local floras (245) for the species of each genus (McLaughlin 2007).

[1]NE, northeastern Arizona; CH, Central Highlands; CA, central Arizona; SW, southwestern Arizona; SE, southeastern Arizona

includes *Asanthus* and *Kuhnia*. *Opuntia s.l.* includes Arizona species of *Cylindropuntia* and *Grusonia*. *Senecio s.l.* includes recently segregated genera *Packera, Barkleyanthus,* and *Ligularia. Gilia s.s.* excludes *Ipomopsis* and *Allophyllum. Oenothera s.s.* excludes *Calylophus* and *Camissonia. Potentilla s.s.* excludes *Argentina, Dasiphora, Ivesia,* and *Sibbaldia*.

There is considerable variation in the mean ranges among the genera that are most speciose in the Arizona flora. Genera with the most widespread species are *Juncus, Salix, Ranunculus, Potentilla, Carex,* and *Cyperus*. The species of these genera are most often found in moist habitats. On the other hand, the narrowly distributed species, including those of *Phacelia, Astragalus, Camissonia, Mentzelia, Gilia,* and *Eriogonum,* are mostly found in drier upland habitats. Species in the genus *Penstemon* are also narrowly distributed, but they occur in a broader range of habitats.

Floristic Areas of Arizona

The five geographic areas shown in table 7.3 are delineated in figure 7.1 to reflect the major floristic subdivisions of the state. These geographic areas are northeastern Arizona, the Central Highlands, central Arizona, southwestern Arizona, and southeastern Arizona.

Northeastern Arizona includes most of the Colorado Plateau, excluding the Kaibab Plateau and San Francisco Peaks. It is an area of largely cold deserts dominated by shrubs, particularly species of *Atriplex, Artemisia,* and *Chrysothamnus,* at the lower elevations and pinyon-juniper woodlands at the higher elevations. Large genera that are most speciose in northeastern Arizona include *Astragalus, Eriogonum, Phacelia, Penstemon,* and *Cryptantha,* all genera with small mean range sizes.

The Central Highlands extend from the White Mountains of east-central Arizona along the Mogollon Rim in the central part of the state to the San Francisco Peaks and the Kaibab Plateau of the north-central part of the state. This floristic subdivision is an area of nearly continuous coniferous forests with scattered montane grasslands. As indicated in table 7.3, the large genera that reach their peak

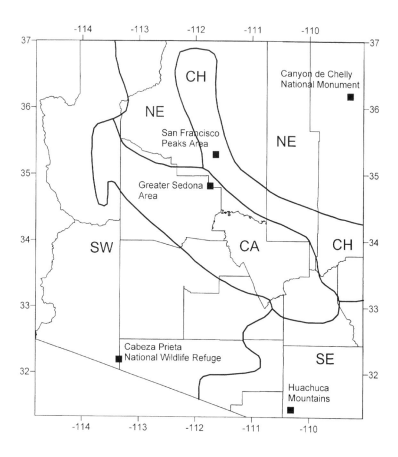

Figure 7.1. The major floristic subdivisions of the state. NE, northeastern Arizona; CH, Central Highlands; CA, central Arizona; SW, southwestern Arizona; SE, southeastern Arizona.

in terms of species richness include *Carex, Senecio, Ranunculus, Potentilla,* and *Salix.* Species of *Erigeron* and *Juncus* are also well represented in the flora. These characteristic genera are among the most widespread of the large genera found in Arizona.

Central Arizona is a large area of mostly deserts, grasslands, interior chaparral communities, and oak and pinyon-juniper woodlands. The region stretches across the central part of the state south of the Mogollon Rim and Colorado Plateau from the Hualapai Mountains in Mohave County in the northwest to the Santa Teresa Mountains of northwestern Graham County in the southeast. Many of the genera listed in table 7.3 are well represented in this floristic subdivision, but only *Lupinus* and *Oenothera* have their maximum species richness here. *Eriogonum, Euphorbia, Brickellia, Juncus,* and *Salix* also have large numbers of species.

Southwestern Arizona encompasses the Sonoran and Mojave Deserts from western Mohave County throughout the southwestern portion of the state. While the two deserts differ in some of their more conspicuous dominants and subdominants—for example, *Carnegia gigantea* (saguaro), *Olneya tesota* (desert ironwood), and species of *Cercidium* (paloverde) are commonly found in the Sonoran Desert and *Yucca brevifolia* (Joshua tree) and *Y. schidigera* (Mojave yucca) in the Mojave Desert—the majority of species inhabiting these deserts are suffrutescent perennials (having a stem that is woody only at the base), herbaceous perennials, and winter annuals. Plants of both deserts consist mostly of warm desert shrubs, such as *Larrea tridentata* (creosote bush) and species of *Ambrosia* (bursage). Grasslands and woodlands are situated at higher elevations. *Opuntia, Gilia,* and *Camissonia* reach their maximum richness in this area. *Phacelia, Cryptantha, Mentzelia,* and *Atriplex* also have high species richness here.

Southeastern Arizona includes all of the sky-island region in the southeastern corner of the state. This floristic subdivision is mostly an area of deserts, grasslands, and oak and pine-oak woodlands, with conifer forests on the higher mountain ranges including the Santa Catalinas, Rincons, Santa Ritas, Pinaleños, and Chiricahuas. Large genera with high species richness include *Muhlenbergia, Euphorbia, Erigeron, Dalea, Cyperus, Asclepias,* and *Brickellia.*

Floristic areas are usually considered in a hierarchy of kingdoms, regions, provinces, subprovinces, and districts (Takhtajan 1986; McLaughlin 1992). All of Arizona lies within the Southwestern Floristic Region of North America, which includes parts of three floristic provinces (McLaughlin 2007). Northeastern Arizona is part of the Colorado Plateau Subprovince of the Great Plains Province; southwestern Arizona lies within the Mojavean Subprovince of the Sonoran Province; and central Arizona and southeastern Arizona lie entirely within the Apachian Subprovince of the Madrean Province. On a finer scale, it is possible to subdivide the Apachian Subprovince into two floristic districts corresponding to central Arizona and southeastern Arizona (McLaughlin 1992). The Central Highlands region constitutes a broad transition zone between the Great Plains and Madrean floristic provinces.

Floristic Affinities and Floristic Elements

The term *floristic affinities* refers to shared taxa between floristic areas. Groups of taxa with broadly overlapping geographic ranges are also called *floristic elements*. Using quantitative analyses of distributions of species, genera, and families in the data set for 245 North American local floras, I have defined floristic elements on maps that illustrate their geographic distributions (McLaughlin 2007). These maps make it possible to identify the floristic elements that contribute the most to the vascular flora of a region and where those elements are best represented within the region.

Distributions of the eight subprovincial floristic elements that are important in Arizona are shown by a series of maps in figure 7.2. The delineation of these elements is based on analyses of presence or absence of individual species in the data set (McLaughlin 2007). These maps, therefore, show broadly overlapping groups of species whose centers of distribution define distinctive floristic subprovinces. The isolines illustrated on the maps are loadings from principal component analyses, which can be interpreted as showing the similarity between floristic elements and local floras used to define those elements.

The Colorado Plateau element extends into northeastern Arizona from its center in southeastern Utah (fig. 7.2A). It is poorly represented below the Mogollon Rim. Taxa that contribute the most to this element are species of cold-desert communities and include many edaphic endemics (plants found only on particular soil types

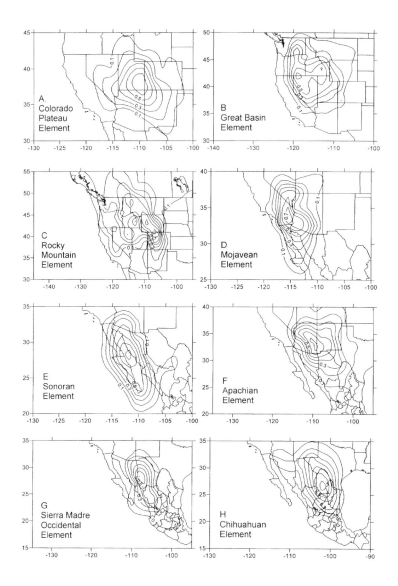

Figure 7.2. Eight selected subprovincial floristic elements of North America.

or geological substrates). Some authorities combine the Great Basin and Colorado Plateau into an Intermountain Region (Reveal 1979). While there is a Great Basin element (fig. 7.2B), it is distinct from the Colorado Plateau element. This latter element is largely restricted to the Arizona Strip area north of the Grand Canyon in the southwestern part of the Colorado Plateau.

The Rocky Mountain element barely enters the eastern part of the Central Highlands of Arizona from New Mexico (fig. 7.2C). An extension of this element across the Central Highlands is not evident in figure 7.2C because the author's data set did not include local floras from the Central Highlands area. The affinities of this area are explored further when relationships of the San Francisco Peaks flora are discussed below. It is evident that the Rocky Mountain element is not especially well represented in southeastern Arizona. Two local floras from southeastern Arizona—those of the Pinaleño Mountains and the Huachuca Mountains—were included in the data set used to map the floristic elements.

The dominant floristic element in southwestern Arizona is the Mojavean element centered in southeastern California and extending into warm desert areas of both northwestern and southwestern Arizona (fig. 7.2D). This element extends over both the Sonoran and Mojave Deserts. A Sonoran element (fig. 7.2E) also extends into southwestern Arizona from northwestern Mexico. The Sonoran element is comprised of species flowering in the summer and early fall in response to precipitation in the summer monsoon season, while the Mojavean element is composed mostly of species flowering in the spring in response to cold-season precipitation.

The Apachian element is centered in southeastern Arizona extending northwest into central Arizona, east into southwestern New Mexico, and south into northeastern Sonora and northwestern Chihuahua (fig. 7.2F). The Apachian element is distinguishable from the Sierra Madre Occidental element entering southeastern Arizona from the Sierra Madre Occidental of northern Mexico (fig. 7.2G). A Chihuahuan element centered in Coahuila also enters southeastern Arizona from southwestern New Mexico and northern Chihuahua (fig. 7.2H). Species associated with the Chihuahuan element are most commonly found on limestone and other calcareous soils in southeastern Arizona.

While the floras of central Arizona and southeastern Arizona are both dominated by species associated with the Apachian element, they differ in several respects. Neither the Sierra Madre Occidental element nor the Chihuahuan element extend much beyond southeastern Arizona into central Arizona. A Californian element including *Fremontodendron californicum* (California flannelbush), *Garrya flavescens* (ashy silktassel), and *Rhus ovata* (sugar sumac)—not shown in figure 7.2—is also recognizable in the flora of central Arizona but is largely absent in the flora of southeastern Arizona.

Another way to examine floristic affinities is mapping the similarities of individual local floras. I selected one well-documented local flora from each of the five floristic subdivisions shown in figure 7.1 for such an examination. These are the Canyon de Chelly National Monument (fig. 7.3) in Apache County (Rink 2005) for northeastern Arizona; the San Francisco Peaks area (fig. 7.4) in Coconino County (Moir 2006) for the Central Highlands; the Greater Sedona area (fig. 7.5) of Coconino and Yavapai Counties (Licher 2003) for central Arizona; the Cabeza Prieta National Wildlife Refuge (fig. 7.6) in Pima and Yuma Counties (Felger 1998) for southwestern Arizona; and the Huachuca Mountains (fig. 7.7) in Cochise County (Bowers and McLaughlin 1996) for southeastern Arizona. The isolines in figures 7.3 to 7.7 are based on similarities using the Otsuka's Index between the local flora and local floras from the rest of North America.

The flora of Canyon de Chelly National Monument includes 647 native species with an average range of 30.3 floras/species out of the data set of 245 local floras from North America (McLaughlin 2007). Canyon de Chelly National Monument shows high affinities throughout the Colorado Plateau of Arizona, Utah, Colorado, and New Mexico (fig. 7.3). Affinities with the western Great Plains region are evident. While Canyon de Chelly is located in the Intermountain Region, its flora has a closer relationship to that of the Great Plains than to that of the Great Basin. There is also a distinct affinity with the floras surrounding the Rio Grande Valley to the southeast.

The San Francisco Peaks flora includes 689 native species with an

Figures 7.3 through 7.7 depict floristic affinities between various local Arizona floras and those from the rest of North America. The isolines represent similarities using the Otsuka's Index.

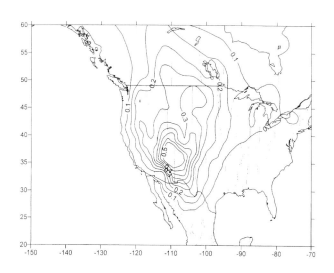

Figure 7.3. Canyon de Chelly National Monument.

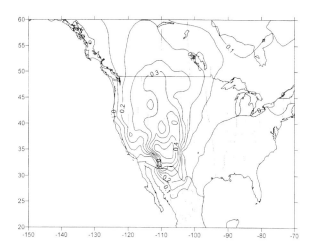

Figure 7.4. The San Francisco Peaks area.

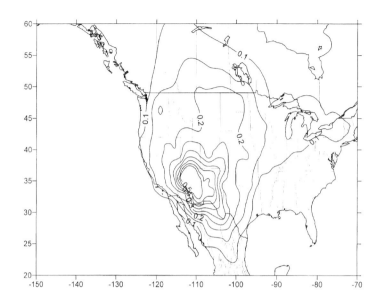

Figure 7.5. The Greater Sedona area.

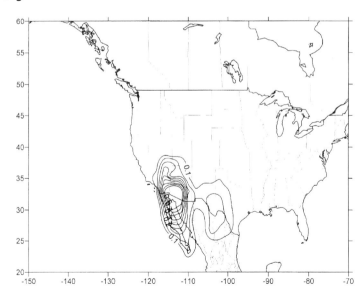

Figure 7.6. The Cabeza Prieta National Wildlife Refuge.

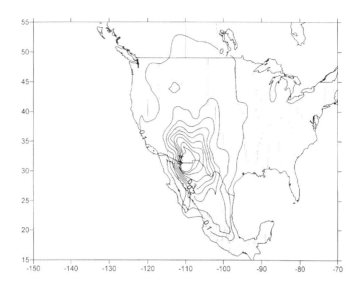

Figure 7.7. The Huachuca Mountains.

average range of 28.9 floras/species (fig. 7.4), demonstrating clear affinities of the Central Highlands with the southern Rocky Mountains of New Mexico and Colorado that were not evident in figure 7.2C. Affinities of the San Francisco Peaks flora extend up through the Rocky Mountain cordillera to southern Canada.

The Greater Sedona area, with 915 native species averaging 27.5 floras/species, covers a large area of southern Coconino County and adjacent Yavapai County. This flora shows high affinities throughout the state except for the southwestern corner (fig. 7.5). These affinities of this flora extend much further to the north, east, and southeast than to the west and southwest.

Cabeza Prieta National Wildlife Refuge covers the largest area of the five local floras considered, but it has the smallest flora of 360 native species with the narrowest average range of only 15.4 floras/species. Affinities extend into adjacent Sonora to the south, southeastern California to the west, and the Mojave Desert of southern Nevada to the north (fig. 7.6). There is also evidence of affinities

with the northern Chihuahuan Desert of western Texas and northern Mexico. The flora of Cabeza Prieta National Wildlife Refuge
contains few aquatic or wetland species. Absence of these species
from this desert flora contributes to the low average range size of its
component species.

The flora of the Huachuca Mountains includes 913 native species
with an average range of 21.7 floras/species (fig. 7.7). The average
range size, therefore, is larger than that of the flora of Cabeza Prieta
National Wildlife Refuge, but it is much smaller than those in the
floras of Canyon de Chelly National Monument, the San Francisco
Peaks area, and the Greater Sedona area. Affinities of the Huachuca
Mountains flora to the west are low, reflecting the lack of summer
rainfall. While affinities to the north and south appear similar, their
pattern is more complicated. The comparatively high affinities of the
Huachucas with floras from northern Arizona, northern New Mexico, and southeastern Colorado are due mainly to the shared species
of moderate to wide ranges—10 or more of the 245 local floras—
while those with the Sierra Madre Occidental to the south are attributed more to shared species with narrower ranges. At the generic
level, the affinities of southeastern Arizona are much stronger with
the Sierra Madre Occidental to the south than with the Rocky
Mountains to the north (McLaughlin 2007).

Origins of Vascular Flora

Any discussion of the origins of a flora is based on the assumption
that speciation rates are not random with respect to geography and
Earth history. Both fossil records (Axelrod and Raven 1985) and the
present distributions of organisms suggest the times and places of
origin of various elements of modern vascular floras. However, there
is little definitive information on the origins of Arizona's vascular
flora. Unfortunately, the state has a weak Neogene fossil record of
vascular plants (Graham 1999). While there are comparatively good
late Pleistocene and Holocene fossil records, particularly macrofos-
sils from woodrat middens and pollen, the conventional viewpoint
seems to be that most of the species of the modern flora of North

America had evolved before the late Pleistocene (Tidwell et al. 1972). The available woodrat midden record has been interpreted mainly as a record of migration of taxa rather than of speciation and extinction (Van Devender et al. 1987).

Species of any regional flora either originated within that region (autochthonous) or migrated into it from outside the region (allochthonous). Biogeographers often conclude that the majority of species within a regional flora have an allochthonous origin. For example, Reveal (1979) suggested that most of the species currently found on the Colorado Plateau and in the Great Basin originated outside of the Intermountain Region. The maps of modern floristic elements and floristic affinities presented in this chapter provide some indication of where species might have originated and the migration routes by which they entered Arizona.

Much of the vascular flora of the Central Highlands (fig. 7.4) and, to a lesser extent, that of the high elevations of southeastern Arizona (fig. 7.7) is shared with the Rocky Mountains (fig. 7.2C). The Rocky Mountain element in Arizona consists mostly of mesophytic species with relatively wide distributions. This element was likely more extensive during glacial periods of the Pleistocene, which accounted for more than 90 percent of that epoch. This element, therefore, can be considered a relictual element in Arizona in the current interglacial period, that is, in the Holocene.

Species adapted to drier conditions probably have varying origins. The southern Great Plains has probably been a source of grassland taxa migrating from the east into northeastern Arizona (fig. 7.3), Central Arizona (fig. 7.5), and southeastern Arizona (fig. 7.7). Warm desert plants migrated into Arizona from the south as part of the Sonoran (fig. 7.2E) and Chihuahuan elements (fig. 7.2H). Many of the taxa of the deserts, grasslands, and oak and Mexican pine-oak woodlands of southeastern Arizona must have originated in the Sierra Madre Occidental of northern Mexico (fig. 7.2G).

Knowledge of the frequency distribution of ranges of Arizona's vascular plant species (fig. 7.8) is also relevant to a discussion of the origins of the state's vascular flora. The x-axis in figure 7.8 represents the number of local floras of the 245 from North America in which a species might occur (McLaughlin 2007). The *y*-axis is the number of species found in these local floras plotted in both an arithmetic scale (fig. 7.8A) and a log scale (fig. 7.8B) to illustrate the frequency distribution of widespread species. This frequency distribution is skewed toward species with narrow ranges. The mean range size for species in the state is 15.5 floras/species, although 70 percent of the species fall below that mean. The median—a more meaningful measure of central tendency in this case—is only 7.5 floras/species. The majority of Arizona's plant species, therefore, are narrowly distributed at the continental scale.

It is difficult to draw conclusions about the origins of species with wide distributional ranges. These species include herbaceous aquatic and wetland plants, ruderal species inhabiting disturbed sites, and woody species that dominate the modern vegetation. Plant species recorded in the fossil record also tend to have comparatively wide ranges. Species occurring in 30 or more local floras or about the average range size of species in many local floras of the state, such as Canyon de Chelly National Monument, San Francisco Peaks, and Greater Sedona area, constitute a small proportion (15 percent) of the total flora of the state.

It is probably reasonable to conclude that the state's vascular flora is largely allochthonous because of the low number (less than 5 percent) of endemic species in the Arizona flora. Consistent with this conclusion is the fact that no floristic province or subprovince lies wholly within the state boundaries (McLaughlin 2007). That is not to say, however, that a majority of Arizona's plant species migrated into the state from distant geographic origins.

The pattern for the frequency distribution of range sizes for Arizona's vascular plants—a large number of narrowly distributed species and a smaller number of widely distributed ones—is a general frequency distribution at regional, continental, and global scales. However, it is a poorly understood frequency distribution with several implications. A small proportion of the species with narrow ranges are likely relics or old taxa that were formerly more widely distributed. Many species with narrow ranges might be restricted in their tolerances to a limited range of climatic and/or edaphic condi-

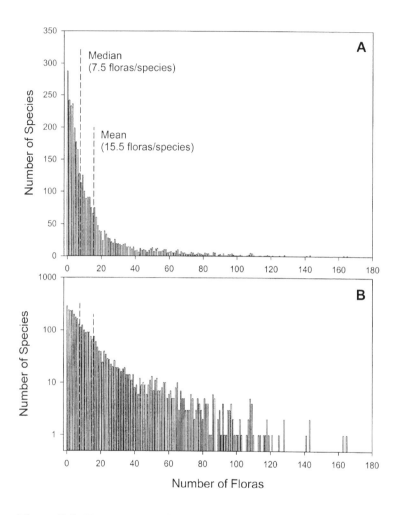

Figure 7.8. Frequency distribution of ranges of native plant species in Arizona's flora with the number of species plotted in (A) an arithmetic scale and (B) a log scale.

tions. Another hypothesis is that narrowly distributed species have broader tolerances than suggested by their current ranges but they have not had sufficient time to fully occupy their potential ranges. In the latter case, the majority of species with narrow ranges are not likely to occur far from the sites in which they originated and are likely to have originated in the not-too-distant past.

The Colorado Plateau, the sky-islands region of southeastern Arizona, and the Sonoran Desert are regions with arid and semiarid climates and large numbers of narrowly distributed species, particularly those in large speciose genera. The arid climates in western North America evolved relatively recently, in the late Neogene (Graham 1999). Therefore, areas of aridity and narrowly distributed plant species in speciose genera are likely to be regions of relatively recent speciation and diversification of the flora.

Most of the plant species with narrow ranges in Arizona also occur in the adjacent states. However, the state boundaries are political in nature and do not coincide with major biogeographic barriers. It might eventually be possible, for instance, to determine whether a narrowly distributed species of the Colorado Plateau originated in Arizona and spread into Utah, New Mexico, or Colorado or originated in one of these adjacent states and then spread into Arizona. In any case, the autochthonous element in Arizona's flora is likely to be higher than that represented by the 5 percent of the extant flora that is strictly endemic to the state.

Concluding Comments

Biogeographic analyses based on phylogenetic studies should eventually provide new insight into the origins of biotas (Lomolino et al. 2006). The actual focus of such studies is placed on clades, that is, branches of the phylogenetic tree. Biotas are composed of hundreds to thousands of such clades. Phylogenetic systematics is a relatively recent field of study, and the intensity of taxon sampling at the species level remains low. The majority of plant species found in Arizona, therefore, have not been included in such studies, and complete phylogenies for the large speciose genera that characterize the vascular flora of the western United States do not exist at this time. When numerous well-supported phylogenies become available, however, it should be possible to identify congruent area-cladograms—substituting information on geographic ranges for their respective species at each terminal branch—that might provide evidence of important historical events not identifiable in the fossil record or inferable from contemporary patterns of distribution.

8
Fauna

PAUL R. KRAUSMAN and PETER F. FFOLLIOTT

Animals are dependent on land and water bodies for their survival, as these landscapes provide the basic habitat components of food, water, and protective cover. Any discussion of fauna, therefore, would be incomplete without mention of the landscapes that provide the necessary elements for their survival and viability. This theme is consistent through distributional descriptions, management practices, conservation programs, and the appreciation of the abundance and distribution of fauna throughout the world. Biologists may never completely understand the mechanisms that cause increases or decreases in the numbers of some species, but we do understand that a species will expire without its basic habitat components. As a result of this situation, many conservation efforts are designed to preserve a species' habitat, especially when biological data about the species are not available. By doing so, the species in question will have the necessary elements available to continue to exist even when biologists do not have a complete knowledge of the species.

Habitat Landscapes

Occurrences and the status of the fauna in Arizona are shaped largely by the diversity of climate, physiography, and vegetation that provides homes for the state's rich array of animal life. Arizona is the only state in the country that contains parts of four desert communities, namely the Sonoran, Chihuahuan, Mojave, and Great Basin deserts, with a climate that is hot and arid. At the opposite end of the spectrum, the state's climate changes to cool and moist in the high elevations of the mostly coniferous forests and treeless alpine tundra on the high peaks of the San Francisco Mountains, making Arizona one of the most diversified areas of wildlife species in the continental United States (Hoffmeister 1986; Corman and Wise-Gervais 2005). Utah, Chihuahua, and Sonora are the only states adjacent to Arizona that have more species of mammals per square mile than Arizona.

A variety of fishes and aquatic and semi-aquatic or transitory herpetofauna, avifauna, mammals, and invertebrates live in or adjacent to the state's streams, rivers, lakes, and reservoirs (fig. 8.1). These bodies of water provide necessary habitat requirements for fish, including oxygen and nutrients, a food supply, and cover. Water, cover, food, and other habitat components are also furnished to other aquatic-dependent animals. Permanent water bodies have the potential to support resident populations of animals, while transitory populations tend to be associated with the more frequently encountered intermittent and ephemeral water bodies in Arizona.

A disproportionate number of rare, threatened, and endangered species inhabit the state's aquatic ecosystems (Johnson 1989; Flather et al. 1994; Baker et al. 2004). Included among the fish species in this dire situation are the humpback chub (*Gila cypha*), Colorado squawfish (*Ptychocheilus lucius*), Little Colorado spinedace (*Lepidomeda vittata*), desert pupfish (*Cyprinodon macularius*), and Gila topminnow (*Poeciliopsis occidentalis*; Miller and Lowe 1964; U.S. Fish and Wildlife Service 1986; Kurpis 2002). Nearly 25 species of nesting birds largely dependent on aquatic habitats have significantly declined in their numbers or been extirpated in the last 100 years (Hunter et al. 1987). Several species of frogs and toads have also declined in numbers or disappeared in recent years (Jennings and Hayes 1994; Stredl and Howland 1995). Listings of endangered and threatened animal and plant species found in Arizona are maintained by the U.S. Fish and Wildlife Service (http://ecos.fws.gov).

One key to sustaining the general well-being of all of the animal and plant species dependent on aquatic habitats in their life cycles is maintaining sufficient amounts of water of appropriate quality and at appropriate times through proper management and careful use of

Figure 8.1. Streams, rivers, lakes, and reservoirs are habitats for a variety of aquatic and semi-aquatic animals in the state.

the state's limited water resources. With continuously increasing demands for more water by urban dwellers and others for various uses, however, Arizona's fauna often become the last in line for obtaining a share of this precious resource.

Terrestrial landscapes are also being challenged by the increasing human population in the state. As the cities, towns, and their surrounding areas attract more people, there will be increasing demands by these residents for the limited natural resources of the state, causing challenges for maintaining the terrestrial habitats of wildlife. Species such as coyotes (*Canis latrans*) and some deer (namely white-tailed deer, *Odocoileus virginianus*; mule deer, *O. hemionus*) populations have readily adapted to the increasing human pressures. However, other species like desert bighorn sheep (*Ovis canadensis*) and mountain lion (*Puma concolor*) do not always fare well with the increasing human population (fig. 8.2). Managers of terrestrial and aquatic habitats must carefully weigh decisions that alter these areas if people and wildlife are to coexist.

Figure 8.2. Mountain lions (*Puma concolor*), such as the male pictured here, fare poorly with the increasing numbers of people in the state.

Fishes

Drastic alterations of aquatic habitats occurred in the state from 1850 to 1960 that significantly upset natural conditions (Lowe 1964b). Water flows in streams and rivers were reduced and some were eliminated, headwaters were impounded, groundwater aquifers were pumped out, and chemical compositions of the water flows were altered, all of which restricted fish habitats. There was also the intentional introduction of exotic fishes that replaced and reduced native species (Minckley 1973). Exotic species have caused competition, predation, and hybridization with native fishes, resulting in a nearly 50 percent decline in the biological diversity of American stream fauna (D'Antonio and Harbensch 1998; Shrader-Frechette

2001; Schade 2003). Of the indigenous fish in Arizona, only the Monkey Springs pupfish (*Cyprinodon arcuatus*) has been eliminated (Minckley and Deacon 1991). However, Apache trout (*Oncorhynchus apache*), humpback chub, Colorado squawfish, Little Colorado spinedace, desert pupfish, and Gila topminnow are among those endangered, while both the number of exotic species and their population sizes continue to increase (Schade and Bonar 2005).

According to Miller (1949), there were 25 native fish species compared to 19 introduced species in Arizona. Lowe (1964b) listed 27 native species and more than 40 exotic fishes in Arizona. Even more exotic fish species were reported in the state's waters by 1973, attributed mostly to accidental and intentional introductions by people (Minckley 1973). Exotic fishes currently dominate the fish species in the state, with Arizona having the highest proportion of nonnative species of fish of 12 western states (Schade 2003; Schade and Bonar 2005). It is likely that nonnative fishes will continue to be a main part of the state's aquatic ecosystems in the future because eradication of nonnative species is usually ineffective and impractical.

People's increasing demands for drinking water and water to irrigate agricultural crops; produce livestock; generate hydropower; maintain lawns, golf courses, and other landscapes; and satisfy numerous other uses are competing with the water resources necessary for sustaining the native fishes of the state. Conservation issues often conflict because much of the public is satisfied with the existence of many of the exotic fish species, as they provide recreational opportunities. However, other people are interested in restoring Arizona's waterways for native species (Rinne and Minckley 1991). It is unlikely that the habitats for native species can be restored because of human interventions that have altered their habitats, for example, changes in water quantity and quality and stream temperature. The state will continue to be a reservoir of exotic fishes as a consequence.

Reptiles and Amphibians

The diversity of reptiles and amphibians is related to the diversity of Arizona's climates and landscapes. Elevations in the state range from about 100 feet on the Lower Colorado River to nearly 12,670 feet atop the San Francisco Peaks, resulting in a range of landscapes with little to abundant rainfall and burning hot to freezing temperatures. More than 12,500 feet of vertical relief in the state, therefore, provides habitat for a diversity of reptiles and amphibians, from the sidewinder (*Crotalus cerastes*) in the deserts surrounding Yuma to the twin-spotted rattlesnake (*Crotalus pricei*) in the Pinaleño Mountains of the southeastern part of the state (Lowe et al. 1986). The herpetofauna of the state includes 132 native species: 27 amphibians and 105 reptiles. One of the amphibians, the Tarahumara frog (*Rana tarahumarae*), has likely been extirpated. The amphibians include one native salamander, the tiger salamander (*Ambystoma tigrinum*), which is widely distributed across North America (Lowe 1964b). The other amphibians are toads and frogs. There are also three introduced amphibian species (Averill-Murray 2003a), the bullfrog (*Rana catesbeiana*), African clawed frog (*Xenopus laevis*), and Rio Grand leopard frog (*Rana berlandieri*).

Six of the reptiles inhabiting the state were introduced (Averill-Murray 2003b), namely the Mediterranean house gecko (*Hemidactylus turcicus*), New Mexico whiptail (*Aspidoscelis neomexicanus*), western spiny-tailed iguana (*Ctenosaura pectinata*), snapping turtle (*Chelydra serpentina*), red-eared slider (*Trachemys scripta*), and spiny soft-shell turtle (*Apalone spinifera*). Native reptiles include 48 lizards, 51 snakes, and six turtles (Averill-Murray 2003b). Thirteen of the reptiles are venomous and potentially dangerous to humans, including the Gila monster (*Heloderma suspectum*), western coral snake (*Micruroides euryxanthus*), and 11 species of rattlesnakes (Lowe et al. 1986).

Increased destruction of the habitats for reptiles and amphibians and increased collection pressure, both legal and illegal, have led to their general protection. For example, the Gila monster has been protected since 1952 and some snakes have been added to the list of protected species. Because venomous reptiles are potentially dangerous to humans and native wildlife, they cannot be imported, exported, transported, propagated, purchased, bartered, sold, leased, or possessed without approval from the Arizona Game and Fish Commission.

Birds

Listings of the birds of Arizona change on a regular basis, as more species move into and out of the state and as biologists learn more about the status of birds. Early listings were prepared by Schwarth (1914), Anderson (1934), Monson and Phillips (1964, 1981), and Phillips et al. (1964), with a more recent accounting compiled by Corman and Wise-Gervais (2005). Regardless of the authority, however, it is generally agreed that Arizona has a rich assemblage of birds inhabiting a diversity of habitat conditions (fig. 8.3). New Mexico is the only state that has more native bird species than Arizona (Monson and Phillips 1981). Due to the diverse climate, physiography, and an abundance of land in public ownership, birders and sportsmen travel to the state from all over the world to observe the diversity of nongame species and enjoy hunting opportunities for game species.

There are at least 307 bird species known to have nested in Arizona (Corman and Wise-Gervais 2005), and at least 475 species of birds are commonly observed throughout the year (Monson and Phillips 1981). The Arizona Bird Committee (T. Corman, member) listed seven exotic bird species in the state. While a complete listing of the common birds of Arizona is not presented in this chapter, those contained in available listings have been verified with a specimen, a clearly recognizable photograph preserved in a scientific collection, or recorded and published information from experienced ornithologists. This diverse array of birds ranges from the common species such as the cactus wren (*Campylorhynchus brunneicapillus*), bushtit (*Psaltriparus minimus*), Mexican jay (*Aphelocoma ultramarina*), juniper titmouse (*Baeolophus ridgwayi*), red-tailed hawk (*Buteo jamaicensis*), turkey vulture (*Cathartes aura*), Gambel's and scaled quail (*Callipepla gambelii* and *C. squamata*), and mourning dove (*Zenaida macroura*) to rare and endangered species. This latter group of species, such as the ferruginous pygmy owl (*Glaucidium brasilianum*) and Mexican spotted owl (*Strix occidentalis lucida*), have created considerable controversy as people's demands for land and resources come into conflict with the birds' habitat requirements.

Figure 8.3. More than 80 species of birds, mammals, reptiles, amphibians, and invertebrates excavate nesting holes or use cavities resulting from the decay of standing trees or holes created by other species in dead or deteriorating trees.

Mammals

Knowledge of Arizona's mammalian fauna is not complete. The first attempt to catalogue the mammals known to reside in the state included 51 species (Coues 1867). Other catalogues have also been published, the most recent by Hoffmeister (1986), who reported 138 indigenous species—other than man—and seven nonnative species of mammals. This number of species is larger than that of any other state in the country, with the exception of California (Cockrum 1960).

Most mammalian species found in Arizona are rarely seen by the casual observer. When they are not noticed, it is largely because of their unique use of their habitats for survival. Many of the mammals are nocturnal (active at night) or reside in crevices, caves, abandoned mines, dense vegetation, rock piles, or underground—precluding casual observation. Furthermore, many of the smaller species, such as bats and smaller rodents, must be "in hand" for their correct identification. With 138 native mammalian species in the state, even the experts cannot identify them all by simply catching a glimpse of the species.

Other mammals are more easily seen with comparatively little effort. These species include lagomorphs such as desert cottontail (*Sylvilagus audubonii*) and jackrabbits (*Lepus* spp.); tree squirrels including Abert's, Kaibab, and Arizona gray squirrel (*Sciurus aberti, S. aberti kaibabensis*, and *S. arizonensis*) and red squirrel (*Tamiasciurus hudsonicus*); and larger mammals such as mule and white-tailed deer (*O. hemionus* and *O. virginianus*), elk (*Cervus elaphus*), and pronghorn (*Antilocapra americana*). That these observations occur is largely because of the animals' comparatively large size and the general openness of the habitats they inhabit. However, some relatively large mammals such as bighorn sheep, mountain lion, and collared peccary (*Pecari tajacu*) are only rarely seen by the casual observer.

Invertebrates

Insects and other invertebrates are the most abundant and diverse fauna inhabiting Arizona's natural environments. While a complete listing of species is too long to present in this chapter, nearly all of the major orders of insects are represented in the state. Equally important to their large numbers, however, are the roles that these invertebrates play in Arizona's natural environments. They are key components in terrestrial and aquatic food webs and often assume crucial functions in the pollination of plants and seed dispersal, nutrient cycling, decomposition activities, and maintaining soil conditions. Some invertebrates are indicators of ecosystem health because they are frequently closely linked to the biotic and abiotic characteristics of their habitats (Kellert 1993; Nelson and Anderson 1994; DeBano and Wooster 2004). Rapid regeneration times and small home ranges of invertebrates such as spiders, butterflies, dragonflies, and ants make them ideal for this purpose. The rapid regeneration times allow biologists to track population changes over time, while small home ranges means that relatively small-scale changes in their habitats such as those associated with management or restoration practices can be evaluated.

Invertebrates are comprised of terrestrial and aquatic components. Terrestrial invertebrates spend their entire life on land, whereas aquatic invertebrates spend at least part of their life in water. Differences in species compositions between terrestrial and aquatic communities in Arizona are often large because of the stark contrasts between the warm and dry uplands and relatively cool and moist aquatic environments. Some invertebrates, such as gall-forming insects, are more abundant and survive at higher rates in terrestrial ecosystems but are poorly adapted to aquatic areas (Fernandes and Price 1997). Other insects, including the ground beetles (Carabidae), are better adapted to living in aquatic habitats that are subject to flooding (Ellis et al. 2001).

Aquatic invertebrate communities are especially species rich. This high richness was illustrated by studies reporting 104 species of invertebrates in Sycamore Creek located in north-central Arizona (Gray 1981) and 63 species inhabiting Aravaipa Creek in the southeastern part of the state (Bruns and Minckley 1980). These numbers of species are comparable to streams in other regions of the country. Aquatic invertebrate communities are also productive in terms of density, biomass, and rate of productivity (Bruns and Minckley 1980; Grimm 1988). Annual rate of invertebrate production in Sycamore Creek has been measured as 135 grams per square meter—the second highest rate of annual productivity at the time of this measurement (Fisher and Gray 1983).

Biodiversity

Efforts have increased recently to sustain the biodiversity of all of Arizona's fauna. Because rare, threatened, and endangered species are the most vulnerable, they receive more attention than the more commonly encountered species, unless those that are common are creating problems for humans. This attention can be illustrated by the situation that pronghorn confront. Arizona provides habitat for three of the five races of pronghorn: the American pronghorn (*Antilocapra americana americana*), Mexican pronghorn (*A. a. mexicana*), and Sonoran pronghorn (*A. a. sonoriensis*). The Sonoran pronghorn has been listed as endangered since 1967. However, necessary recovery efforts were limited until recently because their habitat is inaccessible due to landscape status (wilderness) and military restrictions (Krausman et al. 2005). When their numbers plunged to fewer than 30 individuals statewide in 2002 and the population was literally on the verge of extinction, federal and state agencies initiated aggressive management efforts to reverse the decline. These efforts included forage and water enhancement and captive breeding efforts to increase numbers—concepts that many would not tolerate unless there were no other reasonable options.

Species reintroduction activities are underway in some instances because of people's interest in enhancing the biodiversity of the state's natural environments. As a result, ungulates have been translocated into most of their natural habitats throughout the state. However, predators have not been considered as viable species for translocations until recently. The Mexican gray wolf (*Canis lupus baileyi*), last recorded in Arizona in the 1970s (Nowak 1974), has been protected since 1973. Wolves have been translocated to the southwestern United States in the past 10 years and recovery of this large predator, if successful, will enhance the biodiversity of the state.

Rare, threatened, and endangered species are not the only mammals in which the public is interested. Species that have high population numbers are also cause for concern. Elk in north-central Arizona fit into this category. The number of elk in Arizona has increased significantly in the past few decades, causing monetary losses to some ranchers but enlightenment to the hunting and viewing public. The complete array of Arizona's fauna is enjoyed by many people for a variety of reasons such as their aesthetic values or simply knowing that these species exist in Arizona, even if they are rarely seen. Others ignored them or removed some of them due to damage problems or disease-related issues. However, wildlife does not simply exist for humans, and, therefore, the public needs to realize the important aspects each species plays in maintaining the ecological dynamics of landscapes.

Regardless of their attitudes, people continue to have differing opinions about most types of wildlife. Overall, those views are healthy and people are often excited about maintaining the species that we have, decreasing the numbers of animals that are edging toward extinction, and restoring the species that people have eliminated from parts of their natural habitat.

9
Human Impacts

PETER F. FFOLLIOTT and OWEN K. DAVIS

In this concluding chapter, the editors discuss ways in which Arizona's natural environments have changed in the past 50 years, since the publication of Lowe's book on Arizona's natural environments in 1964. Some of these changes have been the result of natural disturbances, such as changes in climatic patterns, flooding and drought events, or naturally occurring and mostly lightning-caused wildfire. Other changes in these natural environments have been largely attributed to human impacts—both planned and unplanned. While people have little control over the consequences of natural disturbances, they can often mitigate the changes that are largely attributed to their actions.

Some of the more conspicuous human-induced impacts on plant communities, soil and water resources, and wildlife and fish populations that characterize Arizona's natural environments are considered in this chapter. Included are those impacts evolving from timber harvesting operations and other tree-cutting activities, livestock grazing practices, and wildlife-habitat enhancements. Management practices implemented to achieve hoped-for improvements in timber production, rangeland conditions, soil and water resources, and wildlife and fish habitats are part of this discussion, as these practices have resulted in environmental changes in some instances. Wildfire caused by accidental or intentional human ignitions and leading to disruptions of ecological and hydrologic functioning must also be considered. Increasing urbanization, second-home construction, and road-building activities have often altered the integrity of adjacent natural environments and, therefore, must be included. Impacts attributed to these actions are highlighted in a collective sense because the impacts often overlap.

Plant Communities
Forests

Following increases in allowable timber cuttings into the 1960s, when removals were typically one-third to two-thirds of the merchantable volume, timber harvesting levels remained relatively flat into the 1980s. Timber harvesting operations and intensive silvicultural practices leading up to timber harvesting, such as varying thinning treatments, began to decline in the 1990s after a number of lawsuits from environmental organizations challenged most of the timber sales because of the requirement to protect biological diversity and the habitats of rare, threatened, and endangered wildlife and plant species. With these trends in the level of timber harvesting have come the consequent alterations in the structure, stocking, and growth of residual forest stands. Furthermore, the earlier emphasis that managers often placed on maintaining even-aged forest structures has been largely replaced with gradual movement to more natural, uneven-aged structures.

Increases in the number of large human-ignited wildfires also altered forest compositions, structures, and growth patterns. Forest ecosystems experiencing high-severity fire either were destroyed or their ecological and hydrologic functioning were disrupted by the burning, while some of the functioning of forests burned at lower severities is restored in a few years. Compared to lower severity fire, high-severity fire causes a greater loss of vegetative cover, which would protect a site from erosion, and vegetation recovers more quickly on a site experiencing a lower severity fire. In addition, high-severity fires often create water-repellent soil that causes a decrease in infiltration of water into the soil and an increase in surface runoff. The effects of soil water repellency diminish more rapidly on sites burned at lower severities.

Such alterations were observed following the Rodeo-Chediski Wildfire of 2002, the largest wildfire in Arizona's history (Ice et al.

Figure 9.1. Many of the ponderosa pine (*Pinus ponderosa*) trees in stands that burned at a high severity in the Rodeo-Chediski Wildfire of 2002 were killed or severely damaged by the fire.

2004; Neary et al. 2005). Most of the trees in the overstories of ponderosa pine stands that burned at a high severity in this 475,000-acre fire were killed outright or severely damaged and died within a few years of the fire (fig. 9.1). However, many of the stands that burned at lower severities remained intact in their structure, although some of the smaller, lower trees burned. A mosaic of intermingling burned openings within unburned forest overstories was created as a consequence of these burning patterns. Such a spatial pattern is also found in other forests of the state that have experienced large wildfires in the past 50 years.

One consequence of increasing urbanization and the expanding construction of second homes in Arizona has been the expansion of urban-forest interfaces throughout the state. Urbanization and home-building in forest settings has encroached up to the boundaries of mainly public forests in many areas. One consequence of these anthropological developments has been a constant threat of wildfire starting in the forests and spreading into these infrastructures, which has had dire consequences when it occurs.

Species compositions and the production of grasses, forbs, and shrubs in the herbaceous understories of forest communities often change when tree overstories are reduced or tree overstories are removed by timber harvesting operations, vegetative conversion treatments, or wildfire occurrences. The magnitudes of these impacts on herbaceous plants are site-specific and largely dependent on the type of the timber harvesting operation, such as clear-cutting, partial clearing, or selection cutting; the method of converting from one vegetation type to another, including mechanical removal, application of herbicides, or prescribed fire; or the intensity, severity, and duration of a wildfire. An increase in the production of herbaceous plants does not necessarily relate to a corresponding increase in forage plants for livestock, deer, or elk, however. Some of the herbaceous plants are valuable forage for ungulates, other herbaceous plants have forage value to only one group of herbivores, and still other plants are nonforage species in that they have little dietary value. It is often the case that plants of lesser forage value quickly inhabit the disturbed sites. Invasive plants such as the common mullein (*Verbascum thapsus*) are also observed on these sites.

Oak and Pinyon-Juniper Woodlands and Interior Chaparral Communities

Many of the trees and shrubs in oak and pinyon-juniper woodlands and interior chaparral communities have been eliminated or their numbers reduced by mechanical removals, applications of herbicides, or prescribed burning treatments to enhance rangeland productivity and/or improve water yields from the treated areas (De-Bano 1999). Most of these often-extensive vegetative manipulations were discontinued in the 1980s because of environmental concerns. However, scattered cuttings of trees for firewood, poles, and posts have continued in the oak and pinyon-juniper woodlands, although the number of trees cut has significantly declined since the early 1990s.

Human-ignited wildfires have also occurred in these plant communities and, as a result, eliminated overstory shrubs or trees on severely burned sites and reduced their densities on sites burned at lower severities. The intensities, severities, and impacts of these wildfires on natural resources are generally less than those observed in forest communities, however. Fuel buildups are smaller in these plant communities, and the containment of fire is easier.

Urban interfaces have also been expanding outward into oak and pinyon-juniper woodlands and interior chaparral communities. Wildfires starting in these plant communities and spreading into these developments remain a constant concern.

Increasing herbage production and changing herbaceous plant compositions have often occurred with elimination of the overstories or reduction in overstory densities in the pinyon-juniper woodlands and interior chaparral communities. However, the situation is less clear in oak woodlands. Insignificant overstory-understory relationships have been reported in oak woodlands of the southeastern part of the state (Gottfried and Ffolliott 2002; Ffolliott and Gottfried 2005). A similar finding was observed in the Gambel oak (*Quercus gambelii*) stands that intermingle with ponderosa pine forests in the northern part of the state (Reynolds et al. 1970). It might be that removals of oak trees have little impact on the associated herbaceous understories.

Deserts and Grasslands

Invasions of mesquite and other small trees and shrubs into the desert and grassland communities of Arizona has been a common occurrence since the early 1900s. Overgrazing of livestock up to the 1930s, fire-suppression policies of management agencies, and frequently encountered drought conditions are often cited collectively as the main casual factors for these invasions (McPherson 1995; McClaran et al. 2003). A general decline in herbage production has resulted from these invasions as well. Many forage plants have been eliminated or had their levels of production reduced. At the same time, the spread of the exotics Lehmann lovegrass (*Eragrostis lehmanniana*) and buffelgrass (*Pennisetum ciliare*) and other noxious plant species has increased, displacing native species.

There have been attempts to restore the impacted rangelands to a higher level of herbage (forage) production by eliminating the competing woody vegetation through controlled burning treatments, applications of herbicides, and mechanical methods (McClaran et al. 2003). However, environmental concerns of the public and regulations of management agencies about these treatments are restricting or prohibiting large-scale use of these control methods, especially those involving applications of herbicides.

The historical impacts of fire in these plant communities is unclear. Early evidence suggested that the communities were mostly free of shrubs and small trees in comparison to the current situation of woody overstories intermingling with herbaceous plants. According to some authorities, the change to this current condition has come about because of an inability of the naturally occurring wildfires at intervals of 5 to 10 years to burn freely (Wright 1980, 1990), which is largely attributed to the fire-suppression policies imposed by the management agencies in the early 1900s. There is interest by some people to allow more natural fire regimes in these and other plant communities in the state.

Desert and grassland communities continue to be impacted by increasing urbanization and expanding subdivision developments in metropolitan areas. These impacts are especially felt in the Phoenix and Tucson metropolitan areas, where natural ecosystems are rapidly replaced by homes, shopping centers, and streets.

Riparian Associations

Varying combinations of wood cutting, conversions for agricultural cultivation, and attempts at regulating water flows by removing riparian (streamside) vegetation impacted many riparian associations in the state up to the 1980s. Large sycamore, willow, and cottonwood trees were eradicated in the process, and only a few of the formerly widespread mesquite bosques remain in riparian corridors at lower elevations (Stromberg et al. 2004). Some of the eliminated trees have been replaced by invasive species such as tamarisk, a phreatophyte that forms dense, almost impenetrable thickets.

Continual pressures by environmental groups and the general public to preserve riparian ecosystems for wildlife habitats, recreational purposes, and aesthetic values resulted in a change in management philosophy in the early 1980s. Eradication of riparian vegetation stopped at this time, and rehabilitating poorly functioning riparian ecosystems to a healthier condition became the primary focus of management (Baker et al. 1999, 2004). Interests in the condition, role, and sustainability of the state's riparian associations has increased as a result.

Soil and Water Resources
Soil Erosion and Sedimentation

Past timber harvesting practices, overgrazing of livestock, and the impacts of high-severity wildfires have been and—with respect to wildfires—remain the primary causes of increased soil erosion on the disturbed sites. To mitigate the accelerated losses, managers have implemented practices to prevent excessive soil loss from occurring in the first place and control soil erosion when it becomes excessive (Tecle et al. 2003). Re-establishing protective plant covers and, where necessary, installing structural controls have been common remedial measures. These efforts have helped to control the rate of soil loss in many of Arizona's natural environments.

Sedimentation is a product of soil erosion. However, only a relatively small portion of the eroded soil passes through and out of a watershed as sediment during most of the normally encountered storm events in the state. Much of the sediment is deposited at the base of slopes, on floodplains, or within stream channels. Sediment concentrations are higher in streamflow from watersheds where forest overstories have been removed or their density reduced by earlier timber harvesting operations and silvicultural practices (Lopes et al. 2001) or uncontrolled wildfire (Gottfried et al. 2003) or other losses of protective plant and litter covers have taken place. Most of the observed high rates of soil erosion and subsequent sediment concentrations peak shortly after the onset of the disturbances and then decline as the impacted site recovers from these disturbance.

Water Quality

Sustaining flows of high-quality water becomes problematic when vegetation on the contributing watersheds has been inappropriately removed, burned by wildfires, or impacted by other actions causing elevated physical, chemical, and biological pollutants in streamflow regimes. Compounding the problem, runoff and seepage from improperly constructed roads often contain high levels of pollutants. To prevent or mitigate these pollution problems and sustain flow of high-quality water, managers implement "best management practices" when these practices are known and where their implementation is feasible (Brown et al. 1993; Brooks et al. 2003). Regulations and amendments to the Clean Water Act of 1972 stipulate that the state must adopt best management practices to control water degradation from point and nonpoint sources of pollution. Mitigating the problems of groundwater contamination is a more difficult task than sustaining high-quality surface water, however, because adequate knowledge about the rates and flow paths of pollutants into groundwater aquifers is lacking.

The Status of Rivers

Arizona's rivers faced a multitude of threats as the 21st century began. The dams, reservoirs, and water diversions built through the 1960s to irrigate agricultural crops and provide municipal water supplies have had major influences on the continuity and sustainability of water flowing in large rivers such as the Salt, Verde, and Gila Rivers. Smaller impoundments and diversions of water from smaller rivers and tributary streams such as the Agua Fria, Hassayampa, and Santa Cruz continue to have similar hydrologic effects but on smaller scales. Timing of the flows of water from the smaller tributaries into the larger rivers and, eventually, downstream reservoirs and other points of use have also been altered.

Sections of free-flowing rivers have dried up in some systems because of increased pumping of groundwater to meet people's needs, as has happened on the lower reaches of the Santa Cruz and Upper Verde Rivers. The hydrologic linkages between surface water in rivers and groundwater aquifers have become apparent because of this phenomenon in many of the state's water systems (Wirt and Hjalmarson 2000).

Conjunctive management of surface water and groundwater resources has been implemented in some cases to sustain the water needed to balance ecological and hydrologic functioning of the state's river systems with the expanding economies of the state (Blomquist et al. 2004). Conjunctive management involves sustaining the use of surface water supplies and storage with groundwater supplies and storage. Regulations for the protection of water flows in the rivers and their larger tributaries and the groundwater aquifers that supply water to people are insufficient in many parts of the state—a situation that must change to maintain healthy river systems.

Wildlife and Fish Populations

A variety of mammals, avifauna, herpetofauna, and fishes are listed as threatened, endangered, or rare by the federal and/or state governments. Special management initiatives have been required to insure the survival and well-being of these species. To this end, parts of their natural habitats have been set aside and protected from exploitation to sustain these habitats. However, other parts of their habitats are endangered and vanishing.

Wildlife Populations

Changes in wildlife populations are often less apparent than changes in plant communities or soil and water resources. When the habitat of a wildlife species is detrimentally impacted by management activities or land-use changes, human-ignited wildfires, or urbanization pressures, the species often has to move to other habitats. The inherent mobility of the species is a key to the scale of this movement. Mobility and home ranges of larger species such as deer, elk, and pronghorn are greater than those of smaller species such as tree squirrels, cottontail, and rattlesnakes. The former group of wildlife species, therefore, may be less impacted by local disturbances.

Many of the more severe impacts on wildlife populations are those caused by large-scale wildfires destroying or disrupting the availability of preferred habitat components for a species. Responses of birds, reptiles, amphibians, and mammals inhabiting the state's natural environments to wildfires has been species and site specific. The season of burning; the severity, extent, and uniformity of the fire; the effects of the fire on overstory and understory strata; and the rate of postfire recovery all impact the status of Arizona's wildlife populations.

Ever-expanding urban-wildland interfaces are encroaching onto the habitats of many wildlife populations, often to the detriment of the wildlife. More frequent sightings of deer, coyote, and javelina (peccary) near homes, in urban parks, and along streets are reported each year. Some of these reports include observations of property damage by the wildlife. Not surprisingly, contacts and other interactions between people and wildlife populations are likely to increase with the continuing urbanization.

Fish Populations

Changes in native fish populations are even less obvious than changes in terrestrial wildlife populations, largely because fish are rarely seen in their natural habitats. Nevertheless, many of Arizona's native fish species have declined in their numbers and range in the past 50 years (Rinne 2003). The cumulative effects of dams, reservoirs, and water diversions; introduction of nonnative fish species, such as smallmouth bass (*Micropterus dolomieus*) and green sunfish (*Lepomis cyanellus*); and impacts of varying land uses on the habitats of native fishes have been implicated in this demise. Furthermore, habitat alterations—physical (such as stream-channel characteristics), chemical (including nutrient loadings and temperature), and biological (especially introduced fish species)—have all impacted native fishes.

There was little concern about the effects of wildfire on fishes before the Yellowstone fires of 1988, with this catastrophe generating increased interest in these impacts (Rinne 1996). Recent research efforts, however, are providing a better picture of the effects of wildfire on Arizona's fish populations. While only a few studies have documented direct mortality of fish following wildfire, polluting ash, filling of stream channels with increased postfire sediments, or other indirect effects of fire can be significant factors in fish mortality.

Brown et al. (2001) suggested that wildfire is likely the most significant factor in limiting the long-term viability of the endangered Apache trout. Postfire impacts have also been fatal to some fishes in fire-influenced segments of streams (Rinne 2003). Species have become extirpated from other streams.

Rehabilitation Efforts

With the changes in Arizona's natural environment caused by human impacts have come efforts to rehabilitate the sites that have been disturbed, degraded, or destroyed as a consequence of these impacts. These rehabilitation efforts are site specific and largely dependent on whether the sites were overharvested, overgrazed, or exposed to destructive wildfires. Among the methods of rehabilitation are applications of corrective management practices, mechanical or structural measures, or other interventions that reverse the processes of disturbance, degradation, and destruction and (hopefully) restore ecological and hydrologic functioning of the site. These methods are largely based on long-term studies of the plant communities, soil and water resources, and wildlife and fish populations that have been disrupted by human activities and what can be done to remedy the situation.

Literature Cited

Anderson, A. H. 1934. The Arizona state list since 1914. *Condor* 36:78–83.

Anderson, J. L. 1989. Proterozoic anorogenic granites of the southwestern United States. In: J. P. Jenney and S. J. Reynolds, eds. *Geologic evolution of Arizona*, pp. 211–238. Arizona Geological Society Digest 17, Arizona Geological Society, Phoenix.

Angiosperm Phylogeny Group. 2003. An update of the Angiosperm Phylogeny Group classification for the orders and families of flowering plants: APG II. *Botanical Journal of the Linnean Society* 141:399–436.

Arnold, J. F., D. A. Jameson, and E. H. Reid. 1964. *The pinyon-juniper type in Arizona: Effects of grazing, fire, and tree control.* Production Resources Report 84, U.S. Department of Agriculture, Washington, D.C.

Averill-Murray, R. C. 2003a. *Amphibians of Arizona.* NGB List 019, Arizona Game and Fish Department, Nongame Branch, Phoenix.

Averill-Murray, R. C. 2003b. *Reptiles of Arizona.* NGB List 020, Arizona Game and Fish Department, Nongame Branch, Phoenix.

Axelrod, D. I., and P. H. Raven. 1985. Origins of the Cordilleran flora. *Journal of Biogeography* 12:21–47.

Baker, M. B., Jr., L. F. DeBano, and P. F. Ffolliott. 1999. Changing values of riparian ecosystems. In: M. B. Baker Jr., comp. *History of watershed research in the central Arizona highlands*, pp. 43–47. General Technical Report RMRS-GTR-29, U.S. Department of Agriculture Forest Service, Fort Collins, Colo.

Baker, M. B., Jr., L. F. DeBano, P. F. Ffolliott, and G. J. Gottfried. 1998. Riparian-watershed linkages in the Southwest. In: D. E. Potts, ed. *Rangeland management and water resources*, pp. 347–357. Technical Publication Series TPS-98-1, American Water Resources Association, Herndron, Va.

Baker, M. B., Jr., and P. F. Ffolliott. 1998. *Multiple resource evaluation on the Beaver Creek Watershed: An annotated bibliography (1956–1996).* General Technical Report RMRS-GTR-13, U.S. Department of Agriculture Forest Service, Fort Collins, Colo.

Baker, M. B., Jr., P. F. Ffolliott, L. F. DeBano, and D. G. Neary, eds. 2004. *Riparian areas of the southwestern United States: Hydrology, ecology, and management.* Lewis Publishers, Boca Raton, Fla.

Baumer, M. C., and B. Ben Salem. 1985. The arid zone. In: *Sand dune stabilization, shelterbelts and afforestation in dry zones,* pp. 1–8. FAO Conservation Guide 10, FAO, Rome, Italy.

Beaumont, P. 1993. *Drylands: Environmental management and development.* Routledge, London.

Blomquist, W., E. Schlager, and T. Heikkila. 2004. *Common waters, diverging streams: Linking institutions and water management in Arizona, California, and Colorado.* Resources for the Future, Washington, D.C.

Bowers, J. E., and S. P. McLaughlin. 1996. Flora of the Huachuca Mountains, a botanically rich and historically significant sky island in Cochise County, Arizona. *Journal of the Arizona-Nevada Academy of Science* 29:66–107.

Bowring, S. A., and K. E. Karlstrom. 1990. Growth, stabilization, and reactivation of Proterozoic lithosphere in the southwestern United States. *Geology* 18:1203–1206.

Brooks, K. N., P. F. Ffolliott, H. M. Gregersen, and L. F. DeBano. 2003. *Hydrology and the management of watersheds.* Iowa State University Press, Ames.

Brown, D. E., ed. 1994. *Biotic communities: Southwestern United States and northwestern Mexico.* University of Utah Press, Salt Lake City.

Brown, D. E., and C. H. Lowe. 1980. *Biotic communities of the Southwest.* General Technical Report RM-78, U.S. Department of Agriculture Forest Service, Fort Collins, Colo.

Brown, D. K., A. A. Echelle, D. L. Propst, J. E. Brooks, and W. L. Fisher. 2001. Catastrophic wildfire and number of populations as factors influencing risk of extinction for Gila trout (*Oncorhynchus gilae*). *Western North American Naturalist* 61:139–148.

Brown, T. C., D. Brown, and D. Brinkley. 1993. Laws and programs for controlling nonpoint source pollution in forest areas. *Water Resources Bulletin* 22:1–13.

Bruns, D. A., and W. L. Minckley. 1980. Distribution and abundance of benthic invertebrates in a Sonoran stream. *Journal of Arid Environments* 3:117–131.

Busch, D. E. 1995. Effects of fire on southwestern riparian plant community structure. *Southwestern Naturalist* 40:259–267.

Campbell, C. J. 1970. Ecological implications of riparian vegetation management. *Journal of Soil and Water Conservation* 25:49–52.

Campbell, C. J., and W. Green. 1968. Perpetual succession of stream channel vegetation in a semiarid region. *Journal of the Arizona Academy of Science* 5:86–98.

Carmichael, R. S., O. D. Knipe, C. P. Pase, and W. W. Brady. 1978. *Arizona chaparral: Plant associations and ecology.* Research Paper RM-202, U.S. Department of Agriculture Forest Service, Fort Collins, Colo.

Cartron, J.-L. E., S. H. Stoleson, P. L. L. Stoleson, and D. W. Shaw. 2000. Riparian area. In: R. L. Jemison and C. Raish, eds. *Livestock management in the American Southwest: Ecology, society, and economics*, pp. 281–327. Elsevier Science, Amsterdam, The Netherlands.

Castor, S. B., and J. E. Faulds. 2001. Post-6 Ma limestone along the southeastern part of the Las Vegas Valley shear zone, southern Nevada. In: R. A. Young and E. E. Spamer, eds. *The Colorado River: Origin and evolution*, pp. 77–79. Grand Canyon Association Monograph 12, Grand Canyon, Ariz.

Cather, S. M., and B. D. Johnson. 1984. *Eocene tectonics and depositional setting of west-central New Mexico and eastern Arizona.* Circular 192, New Mexico Bureau of Mines and Mineral Resources, Socorro.

Chambers, C. L., and R. S. Holthausen. 2000. Montane ecosystems used as rangelands. In: R. Jemison and C. Raish, eds. *Livestock management in the American Southwest: Ecology, society, and economics*, pp. 213–280. Elsevier Science, Amsterdam, The Netherlands.

Chenoweth, M., and C. Landsea. 2004. The San Diego Hurricane of 2 October 1858. *Bulletin of the American Meteorological Society* 85:1689–1697.

Clements, T., R. H. Merriam, R. O. Stone, J. F. Mann Jr., and J. L. Eyman. 1957. *A study of desert surface conditions.* Technical Report EP 53, U.S. Army Quartermaster Research and Development Command, Environmental Protection Research Division, Washington, D.C.

Cockrum, E. L. 1960. *The Recent mammals of Arizona: Their taxonomy and distribution.* University of Arizona Press, Tucson.

Coney, P. J., and S. J. Reynolds. 1977. Cordilleran Benioff zones. *Nature* 270:403–406.

Cooke, R. U., and R. W. Reeves. 1976. *Arroyos and environmental change in the American South-West.* Oxford University Press, New York.

Cooke, R. U., A. Warren, and A. Goudie. 1993. *Desert geomorphology.* University College London Press, London.

Corman, T. E., and C. Wise-Gervais. 2005. *Arizona breeding bird atlas.* University of New Mexico Press, Albuquerque.

Coues, E. 1867. The quadrupeds of Arizona. *American Naturalist* 1:281–292, 351–363, 393–400, 531–541.

Covington, W. W., and M. M. Moore. 1994. Southwestern ponderosa pine forest structure and resource conditions: Changes since Euro-American settlement. *Journal of Forestry* 92(1):39–47.

Cronquist, A. 1988. *The evolution and classification of flowering plants.* New York Botanical Garden, New York.

D'Antonio, C. M., and K. Harbensch. 1998. Community and ecosystem impacts of introduced species. *Fremontia* 26:13–18.

Daubenmire, R. F. 1943. Vegetational zonation in the Rocky Mountains. *Botanical Review* 9:325–393.

Davis, G. A., G. S. Lister, and S. J. Reynolds. 1986. Structural evolution of the Whipple and South Mountains shear zones, southwestern United States. *Geology* 14:7–10.

Davis, G. H. 1980. Structural characteristics of metamorphic core complexes, southern Arizona. In: M. D. Crittenden Jr., P. J. Coney, and G. H. Davis, eds. *Cordilleran metamorphic core complexes*, pp. 35–77. Geological Society of America Memoir 153, Geological Society of America, Boulder, Colo.

DeBano, L. F. 1999. Chaparral shrublands in the southwestern United States. In: P. F. Ffolliott and A. Ortega-Rubio, eds. *Ecology and management of forests, woodlands, and shrublands in the dry regions of the United States and Mexico: Perspectives for the 21st century*, pp. 83–94. Centro de Investigaciones Biológicas del Noroeste, S.C., La Paz, Baja California Sur, Mexico.

DeBano, L. F., and M. B. Baker Jr. 1999. Riparian ecosystems of the southwestern United States. In: P. F. Ffolliott and A. Ortega-Rubio, eds. *Ecology and management of forests, woodlands, and shrublands in the dry regions of the United States and Mexico: Perspectives for the 21st century*, pp. 107–120. Centro de Investigaciones Biológicas del Noroeste, S.C., La Paz, Baja California Sur, Mexico.

DeBano, S. J., and D. E. Wooster. 2004. Insects and other invertebrates: Ecological roles and indicators of riparian and stream health. In: M. B. Baker Jr., P. F. Ffolliott, L. F. DeBano, and Daniel G. Neary. *Riparian areas of the southwestern United States: Hydrology, ecology, and management*, pp. 215–236. Lewis Publishers, Boca Raton, Fla.

Dellenbaugh, L. E. 1932. The Painted Desert. *Science* 76:437.

Dickinson, W. R. 1989. Tectonic setting of Arizona through geologic time. In: J. P. Jenney and S. J. Reynolds, eds. *Geologic evolution of Arizona*, pp.

1–16. Arizona Geological Society Digest 17, Arizona Geological Society, Phoenix.

Dickinson, W. R., and W. S. Snyder. 1978. Plate tectonics of the Laramide orogeny. In: V. Matthews III, ed. *Laramide folding associated with basement block faulting in the western United States*, pp. 355–366. Geological Society of America Memoir 151, Geological Society of America, Boulder, Colo.

Douglas, M. W., R. A. Maddox, K. Howard, and S. Reyes. 1993. The Mexican monsoon. *Journal of the Climate* 6:1665–1677.

Dregne, H. E. 1976. *Soils of the arid regions.* Elsevier, Amsterdam, The Netherlands.

Dregne, H. E. 1983. *Desertification in arid lands.* Harwood Academic Publishers, New York.

DuBois, S. M., and A. W. Smith. 1980. *The 1887 earthquake in San Bernardino Valley, Sonora: Historic accounts and intensity patterns in Arizona.* Special Paper 3, Arizona Bureau of Geology and Mineral Technology, Phoenix.

Eisele, J., and C. E. Isachsen. 2001. Crustal growth in southern Arizona: U-Pb geochronologic and Sm-Nd isotopic evidence for addition of the Paleoproterozoic Cochise block to the Mazatzal province. *American Journal of Science* 301:773–797.

Ellis, L. M., C. S. Crawford, and M. C. Molles. 2001. Influence of annual flooding on terrestrial arthropod assemblages of a Rio Grande riparian forest. *Regulated Rivers: Research and Management* 17:1–20.

Engle, D. M., and T. G. Bidwell. 2000. Plains grasslands. In: R. Jemison and C. Raish, eds. *Livestock management in the American Southwest: Ecology, society, and economics*, pp. 97–152. Elsevier Science, Amsterdam, The Netherlands.

Epple, A. O. 1995. *A field guide to the plants of Arizona.* Globe Pequot Press, Guilford, Conn.

Faulds, J. E., B. C. Schreiber, S. J. Reynolds, D. Okaya, and L. Gonzalez. 1997. Origin and paleogeography of an immense, nonmarine Miocene salt deposit in the Basin and Range (western USA). *Journal of Geology* 105:19–36.

Felger, R. S. 1998. *Checklist of the plants of Cabeza Prieta National Wildlife Refuge, Arizona.* Drylands Institute, Tucson, Ariz.

Fernandes, G. W., and P. W. Price. 1997. The adaptive significance of insect gall distribution: Survivorship of species in xeric and mesic habitats. *Oecologia* 90:4–20.

Ffolliott, P. F. 1999. Encinal woodlands of the southwestern United States. In: P. F. Ffolliott and A. Ortega-Rubio, eds. *Ecology and management of forests, woodlands, and shrublands in the dry regions of the United States and Mexico: Perspectives for the 21st century*, pp. 69–81. Centro de Investigaciones Biológicas del Noroeste, S.C., La Paz, Baja California Sur, Mexico.

Ffolliott, P. F. 2002. Ecology and management of evergreen oak woodlands in Arizona and New Mexico. In: W. J. McShea and W. M. Healy, eds. *Oak forest ecosystems: Ecology and management for wildlife*, pp. 304–316. Johns Hopkins University Press, Baltimore, Md.

Ffolliott, P. F., and M. B. Baker Jr. 1999. Montane forests of the southwestern United States. In: P. F. Ffolliott and A. Ortega-Rubio, eds. *Ecology and management of forests, woodlands, and shrublands in the dry regions of the United States and Mexico: Perspectives for the 21st century*, pp. 39–52. Centro de Investigaciones Biológicas del Noroeste, S.C., La Paz, Baja California Sur, Mexico.

Ffolliott, P. F., and G. J. Gottfried. 1999. Forest formations in the southwestern United States and northwestern Mexico. In: P. F. Ffolliott and A. Ortega-Rubio, eds. *Ecology and management of forests, woodlands, and shrublands in the dryland regions of the United States and Mexico: Perspective for the 21st century*, pp. 7–22. Centro de Invesitgaciones Biological del Norte, S.C., La Paz, Baja California Sur, Mexico.

Ffolliott, P. F., and G. J. Gottfried. 2005. Vegetative characteristics of oak savannas in the southwestern United States. In: G. J. Gottfried, B. S. Gebow, L. G. Eskew, and C. B. Edminster, comp. *Connecting mountain land and desert seas: Biodiversity and management of the Madrean Archipelago II*, pp. 399–402. Proceedings RMRS-P-36, U.S. Department of Agriculture Forest Service, Fort Collins, Colo.

Fisher, S. G., and L. J. Gray. 1983. Secondary production and organic matter processing by collector macroinvertebrates in a desert stream. *Ecology* 64:1217–1224.

Flather, C. H., L. A. Joyce, and C. A. Bloomgarden. 1994. *Species endangerment patterns in the United States*. General Technical Report RM-241, U.S. Department of Agriculture Forest Service, Fort Collins, Colo.

Fletcher, R., and W. A. Robbie. 2004. Historical and current conditions of southwestern grasslands. In: D. M. Finch, ed. *Assessment of grassland ecosystem conditions in the southwestern United States*, Vol. 1, pp. 120–129. General Technical Report RMRS-GTR-135, U.S. Department of Agriculture Forest Service, Fort Collins, Colo.

Ford, P. L., D. U. Potter, R. Pendleton, B. Pendleton, W. A. Robbie, and G. J.

Gottfried. 2004. Southwestern grassland ecology. In: D. M. Finch, ed. *Assessment of grassland ecosystem conditions in the southwestern United States*, Vol. 1, pp. 18–38. General Technical Report RMRS-GTR-135, U.S. Department of Agriculture Forest Service, Fort Collins, Colo.

Gentry, A. H. 1986. Endemism in tropical versus temperate plant communities. In: M. E. Soulé, ed. *Conservation biology: The science of scarcity and diversity*, pp. 153–181. Sinauer Associates, Sunderland, Mass.

Gottfried, G. J. 1992a. Ecology and management of southwestern pinyon-juniper woodlands. In: P. F. Ffolliott, G. J. Gottfried, D. A. Bennett, V. M. Hernandez C., A. Ortega-Rubio, and R. H. Hamre, tech. coord. *Ecology and management of oak and associated woodlands: Perspectives in the southwestern United States and northern Mexico*, pp. 78–86. General Technical Report RM-218, U.S. Department of Agriculture Forest Service, Fort Collins, Colo.

Gottfried, G. J. 1992b. Growth and development of an old-growth Arizona mixed conifer stand following initial harvesting. *Forest Ecology and Management* 54:1–26.

Gottfried, G. J. 1999. Pinyon-juniper woodlands of the southwestern United States. In: P. F. Ffolliott and A. Ortega-Rubio, eds. *Ecology and management of forests, woodlands, and shrublands in the dry regions of the United States and Mexico: Perspectives for the 21st century*, pp. 53–67. Centro de Investigaciones Biológicas del Noroeste, S.C., La Paz, Baja California Sur, Mexico.

Gottfried, G. J., L. F. DeBano, and P. F. Ffolliott. 1999. Creating a basis for watershed management in high-elevation forests. In: M. B. Baker Jr., comp. *History of watershed resources in the central Arizona highlands*, pp. 35–41. General Technical Report RMRS-GTR-29, U.S. Department of Agriculture Forest Service, Fort Collins, Colo.

Gottfried, G. J., and P. F. Ffolliott. 1992. Effects of moderate timber harvesting in an old-growth Arizona mixed conifer watershed. In: M. R. Kaufmann, W. H. Moir, and R. L. Bassett, tech. coord. *Old-growth forests in the Southwest and Rocky Mountain regions: Proceedings of a workshop*, pp. 184–194. General Technical Report RM-213, U.S. Department of Agriculture Forest Service, Fort Collins, Colo.

Gottfried, G. J., and P. F. Ffolliott. 2002. Notes on herbage resources in encinal woodlands. In: W. L. Halvorson and B. S. Gebow, eds. *Meeting resource management information needs*, pp. 53–55. Proceedings of the Fourth Conference on Research and Resource Management in the Southwestern Deserts, Sonoran Desert Field Station, Tucson, Ariz.

Gottfried, G. J., P. F. Ffolliott, and L. F. DeBano. 1995a. Forests and woodlands of the sky islands: Stand characteristics and silvicultural prescriptions. In: L. F. DeBano, P. F. Ffolliott, A. Ortega-Rubio, G. J. Gottfried, R. H. Hamre, and C. B. Edminster, tech. coord. *Biodiversity and management of the Madrean archipelago: The sky islands of the southwestern United States and northwestern Mexico*, pp. 152–164. General Technical Report RM-GTR-264, U.S. Department of Agriculture Forest Service, Fort Collins, Colo.

Gottfried, G. J., D. G. Neary, M. B. Baker Jr., and P. F. Ffolliott. 2003. Impacts of wildfire on hydrologic processes in forest ecosystems: Two case studies. In: K. G. Renard, S. A. McElroy, W. J. Gburek, H. E. Canfield, and R. L. Scott, eds. *First interagency conference on research in watersheds*, pp. 668–673. U.S. Department of Agriculture Agricultural Research Service, Washington, D.C.

Gottfried, G. J., and R. D. Piper. 2000. Pinyon-juniper rangelands. In: R. Jemison and C. Raish, eds. *Livestock management in the American Southwest: Ecology, society, and economics*, pp. 153–211. Elsevier Science, Amsterdam, The Netherlands.

Gottfried, G. J., T. W. Swetnam, C. D. Allen, J. L. Betancourt, and A. L. Chung-MacCoubrey. 1995b. Pinyon-juniper woodlands. In: D. M. Finch and J. A. Tainter, tech. eds. *Ecology, diversity, and sustainability of the Middle Rio Grande Basin*, pp. 95–132. General Technical Report RM-GTR-268, U.S. Department of Agriculture Forest Service, Fort Collins, Colo.

Goudie, A. S. 1996. Climate: past and present. In: W. M. Adams, A. S. Goudie, and A. R. Orme, eds. *The physical geography of Africa*, pp. 34–59. Oxford University Press, New York.

Goudie, A., and J. Wilkinson. 1977. *The warm desert environment*. Cambridge University Press, New York.

Graf, W. L. 1983. The arroyo problem: Palaeohydrology and palaeohydraulics in the short term. In: K. J. Gregory, ed. *Background to palaeohydrology*, pp. 279–302. John Wiley & Sons, New York.

Graham, A. 1999. *Late Cretaceous and Cenozoic history of North American vegetation*. Oxford University Press, New York.

Gray, L. J. 1981. Species composition and life histories of aquatic insects in a lowland Sonoran desert stream. *American Midland Naturalist* 106:229–242.

Green, C. R., and W. D. Sellers. 1964. *Arizona climate*. University of Arizona Press, Tucson.

Grimm, N. B. 1988. Role of macroinvertebrates in nitrogen dynamics of a desert stream. *Ecology* 69:1884–1893.

Hales, J. E., Jr. 1972. Surges of maritime air northward over the Gulf of California. *Monthly Weather Review* 100:298–306.

Hamblin, W. K. 1994. *Late Cenozoic lava dams in the western Grand Canyon.* Geological Society of America Memoir 183, Geological Society of America, Boulder, Colo.

Hart, F. C. 1937. Precipitation and run-off in relation to altitude in the Rocky Mountain region. *Journal of Forestry* 35:1005–1010.

Heald, W. F. 1951. Sky islands of Arizona. *Natural History* 60:56–63, 95–96.

Heathcote, R. L. 1983. *The arid lands: Their use and abuse.* Longman, New York.

Hecht, M. E., and R. W. Reeves. 1981. *The Arizona atlas.* Office of Arid Lands Studies, University of Arizona, Tucson.

Heede, B. H., M. D. Harvey, and J. R. Laird. 1988. Sediment delivery linkages in a chaparral watershed following wildfire. *Environmental Management* 12:349–358.

Hendricks, D. M. 1985. *Arizona soils.* College of Agriculture, University of Arizona, Tucson.

Hendrickson, D. A., and W. L. Minckley. 1984. Cienegas: Vanishing climax communities of the American Southwest. *Desert Plants* 6:131–175.

Henry, C. D., and J. J. Aranda-Gomez. 1992. The real southern Basin and Range: Mid- to late-Cenozoic extension in Mexico. *Geology* 20:701–704.

Hills, E. S., C. D. Ollier, and C. R. Twidale. 1966. Geomorphology. In: E. S. Hills, ed. *Arid Lands, a geographical appraisal,* pp. 53–75. Methuen & Co., Ltd., London, and UNESCO, Paris.

Hoffmeister, D. F. 1986. *Mammals of Arizona.* University of Arizona Press, Tucson.

House, P. K., P. A. Pearthree, J. W. Bell, A. R. Ramelli, and J. E. Faulds. 2002. New stratigraphic evidence for the Late Cenozoic inception and subsequent alluvial history of the lower Colorado River from near Laughlin, Nevada. *Geological Society of America Abstracts* 34(4):A-60.

Hunter, W. C., R. D. Ohmart, and B. W. Anderson. 1987. Status of riparian-obligate birds in southwestern riverine systems. *Western Birds* 18:1–18.

Ice, G. G., D. G. Neary, and P. W. Adams. 2004. Effects of wildfire on soils and watershed processes. *Journal of Forestry* 102(6):16–20.

Jennings, M. R., and M. P. Haynes. 1994. Decline of native ranid frogs in the desert Southwest. In: P. R. Brown and J. W. Wright, eds. *Herpetology of the North American deserts: Proceedings of a symposium,* pp. 183–187. Southwestern Herpetologists Society, Excelsior, Minn.

Johnson, A. S. 1989. The thin green line: Riparian corridors and endangered species in Arizona and New Mexico. In: G. Mackintosh, ed. *Preserving communities and corridors*, pp. 35–46. Defenders of Wildlife, Washington, D.C.

Johnson, R. R., C. D. Ziebell, D. R. Patton, P. F. Ffolliott, and R. H. Hamre, tech. coord. 1985. *Riparian ecosystems and their management: Reconciling conflicting uses.* General Technical Report RM-120, U.S. Department of Agriculture Forest Service, Fort Collins, Colo.

Jones, C. H., L. J. Sonder, and J. R. Unruh. 1998. Lithospheric gravitational potential energy and past orogenesis: Implications for conditions of initial Basin and Range and Laramide deformation. *Geology* 26:639–642.

Jones, J. R. 1974. *Silviculture of southwestern mixed conifer and aspen: The status of our knowledge.* Research Paper RM-122, U.S. Department of Agriculture Forest Service, Fort Collins, Colo.

Jurwitz, L. R. 1953. Arizona's two-season rainfall pattern. *Weatherwise* 6:96–99.

Kartesz, J. T. 1999. A synonymized checklist and atlas with biological attributes for the vascular flora of the United States, Canada, and Greenland. In: J. T. Kartesz and C. A. Meacham, eds. *Synthesis of the North American Flora*, Ver. 1.0. North Carolina Botanical Garden, Chapel Hill.

Kearney, T. H., R. H. Peebles, and collaborators. 1960. *Arizona flora.* University of California Press, Berkeley.

Kellert, S. R. 1993. Values and perceptions of invertebrates. *Conservation Biology* 7:845–855.

Knipe, O. D., C. P. Pase, and R. S. Carmichael. 1979. *Plants of Arizona chaparral.* General Technical Report RM-64, U.S. Department of Agriculture Forest Service, Fort Collins, Colo.

Krantz, R. W. 1989. Laramide structures of Arizona. In: J. P. Jenney and S. J. Reynolds, eds. *Geologic evolution of Arizona*, pp. 463–483. Arizona Geological Society Digest 17, Arizona Geological Society, Phoenix.

Krausman, R. R., J. R. Morgart, L. K. Harris, C. S. O'Brien, J. W. Cain III, and S. S. Rosenstock. 2005. Introduction. In: *Management for the survival of Sonoran pronghorn in the United States.* Wildlife Society Bulletin 33, Wildlife Society, Bethesda, Md.

Kuchler, A. W. 1964. *Potential natural vegetation of the conterminous United States.* Special Publication 36, American Geographical Society of New York, New York.

Kurpis, L. 2002. *Endangered species in Arizona: The rarest info around.* Available via http://www.endangeredspecies.com/states/az.htm

Licher, M. 2003. Flora of the greater Sedona area. Unpublished checklist.

Loftin, S. R., C. E. Bock, J. H. Bock, and S. L. Brantley. 2000. Desert grasslands. In: R. Jemison and C. Raish, eds. *Livestock management in the American Southwest: Ecology, society, and economics*, pp. 53–96. Elsevier Science, Amsterdam, The Netherlands.

Lomolino, M. V., B. R. Riddle, and J. H. Brown. 2006. *Biogeography*. Sinauer Associates, Inc., Sunderland, Mass.

Lopes, V. L., and P. F. Ffolliott. 1993. Sediment rating curves for a clearcut ponderosa pine watershed in northern Arizona. *Water Resources Bulletin* 20:369–382.

Lopes, V. L., P. F. Ffolliott, and M. B. Baker Jr. 2001. Impacts of vegetative practices on suspended sediment from watersheds in Arizona. *Journal of Water Resources Planning and Management* 127:41–47.

Lorenz, D. J., and D. L. Hartmann. 2006. The effect of the MJO on the North American monsoon. *Journal of the Climate* 19:333–343.

Lowe, C. H. 1964a. *Arizona's natural environment*. University of Arizona Press, Tucson.

Lowe, C. H., ed. 1964b. *The vertebrates of Arizona*. University of Arizona Press, Tucson.

Lowe, C. H., C. R. Schwalbe, and T. B. Johnson. 1986. *The venomous reptiles of Arizona*. Arizona Game and Fish Department, Phoenix.

Lucchitta, I. 1979. Late Cenozoic uplift of the southwestern Colorado Plateau and adjacent Lower Colorado River region. *Tectonophysics* 61:63–95.

Lucchitta, I. 1987. The mouth of the Grand Canyon and the edge of the Colorado Plateau in the upper Lake Mead area, Arizona. *Geological Society of America Centennial Field Guide, Rocky Mountain Section* 2:365–370.

Lull, H. W., and L. Ellison. 1950. Precipitation in relation to altitude in central Utah. *Ecology* 31:479–484.

Lynch, D. J. 1989. Neogene volcanism in Arizona: The recognizable volcanoes. In: J. P. Jenney and S. J. Reynolds, eds. *Geologic evolution of Arizona*, pp. 681–700. Arizona Geological Society Digest 17, Arizona Geological Society, Phoenix.

Mabbutt, J. A. 1977. *Desert landforms*. MIT Press, Cambridge, Mass.

Mainguet, M. 1999. *Aridity: Droughts and human development*. Springer-Verlag, New York.

Mann, D. E. 1963. *The politics of water in Arizona*. University of Arizona Press, Tucson.

Mann, M. E., M. A. Cane, S. E. Zebiak, and A. Clement. 2005. Volcanic and solar forcing of El Niño over the past 1000 years. *Journal of the Climate* 18:447–456.

Martin, P. S. 1963. *The last 10,000 years: A fossil pollen record of the American Southwest.* University of Arizona Press, Tucson.

McClaran, M. P., P. F. Ffolliott, and C. B. Edminster, tech. coord. 2003. *Santa Rita Experimental Range: 100 years (1903–2003) of accomplishments and contributions.* Rocky Mountain Research Station Proceedings RMRS-P-30, U.S. Department of Agriculture Forest Service, Fort Collins, Colo.

McClaran, M. P., and G. R. McPherson. 1999. Oak savannas in the American Southwest. In: R. C. Anderson, J. S. Fralish, and J. M. Basin, eds. *Savannas, barrens, and rock outcrop plant communities of North America*, pp. 275–287. Cambridge University Press, New York.

McClaran, M. P., and T. R. Van Devender, eds. 1995. *The desert grassland.* University of Arizona Press, Tucson.

McDonald, J. E. 1956. *Variability of precipitation in an arid region: A survey of characteristics for Arizona.* Technical Report 1, Institute of Atmospheric Physics, University of Arizona, Tucson.

McGinnies, W. G. 1968. Vegetation of desert environments. In: W. G. McGinnies, B. J. Goldman, and P. Paylore, eds. *Deserts of the world*, pp. 379–566. University of Arizona Press, Tucson.

McGinnies, W. G., B. J. Goldman, and P. Paylore, eds. 1968. *Deserts of the world: An appraisal of research into their physical and biological environments.* University of Arizona Press, Tucson.

McHenry, D. E. 1934. Canadian zone plants on the south rim of the Grand Canyon. *Grand Canyon Nature Notes* 9:301–302.

McLaughlin, S. P. 1992. Are floristic areas hierarchically arranged? *Journal of Biogeography* 19:21–32.

McLaughlin, S. P. 2004. Riparian flora. In: M. B. Baker Jr., P. F. Ffolliott, L. F. DeBano, and D. G. Neary, eds. *Riparian areas of the southwestern United States: Hydrology, ecology, and management*, pp. 127–167. Lewis Publishers, Boca Raton, Fla.

McLaughlin, S. P. 2007. Tundra to tropics: The floristic plant geography of North America. *Sida Botanical Miscellany* 30:1–58.

McPherson, G. R. 1992. Ecology of oak woodlands in Arizona. In: P. F. Ffolliott, G. J. Gottfried, D. A. Bennett, V. M. Hernandez C., A. Ortega-Rubio, and R. H. Hamre, tech. coord. *Ecology and management of oak and associated woodlands: Perspectives in the southwestern United States and northern Mexico*, pp. 24–33. General Technical Report RM-218, U.S. Department of Agriculture Forest Service, Fort Collins, Colo.

McPherson, G. R. 1995. The role of fire in the desert grassland. In: M. P. McClaran and T. R. Van Devender, eds. *The desert grassland*, pp. 130–151. University of Arizona Press, Tucson.

McPherson, G. R. 1997. *Ecology and management of North American savannas.* University of Arizona Press, Tucson.

Mead, P. 1930. A brief ecological comparison of life-zones on the Kaibab Plateau. *Grand Canyon Nature Notes* 5:13–17.

Medina, A. L. 1986. Riparian plant communities of the Fort Bayard watershed in southwestern New Mexico. *Southwestern Naturalist* 31:345–359.

Meigs, P. 1953. World distribution of arid and semi-arid homoclimates. *Arid Zone Hydrology: UNESCO Arid Zone Research Series* 1:203–209.

Meigs, P. 1973. World distribution of coastal deserts. In: D. H. K. Amiran and A. W. Wilson, eds. *Coastal deserts: Their natural and human environments*, pp. 2–12. University of Arizona Press, Tucson.

Menges, C. M., and P. A. Pearthree. 1989. Late Cenozoic tectonism in Arizona and its impact on regional landscape evolution. In: J. P. Jenney and S. J. Reynolds, eds. *Geologic evolution of Arizona*, pp. 649–680. Arizona Geological Society Digest 17, Arizona Geological Society, Phoenix.

Merriam, C. H. 1891. Results of a biological reconnaissance of south-central Idaho. In: *North American Fauna*, Vol. 5, pp. 1–127. U.S. Department of Agriculture, Division of Ornithology and Mammalogy, Washington, D.C.

Merriam, C. H. 1898. *Life-zones and crop-zones of the United States.* Biological Survey Bulletin 10, U.S. Department of Agriculture, Washington, D.C.

Merriam, C. H., and L. Steineger. 1890. Results of a biological survey of the San Francisco Mountains region and desert of the Little Colorado in Arizona. In: *North American Fauna*, Vol. 3. U.S. Department of Agriculture, Division of Ornithology and Mammalogy, Washington, D.C.

Metzger, D. G., and O. J. Loeltz. 1973. *Geohydrology of the Needles area, Arizona, California, and Nevada.* Professional Paper 486-J, U.S. Geological Survey, Washington, D.C.

Metzger, D. G., O. J. Loeltz, and B. Irelan. 1973. *Geohydrology of the Parker-Blythe-Cibola area, Arizona and California.* Professional Paper 486-G, U.S. Geological Survey, Washington, D.C.

Middleton, N., and D. Thomas, eds. 1997. *World atlas of desertification.* United Nations Environment Programme and Arnold, London.

Miller, R. R. 1949. *Keys for the identification of the fishes of Arizona.* Museum of Zoology, University of Michigan, Ann Arbor.

Miller, R. R., and C. H. Lowe. 1964. Fishes of Arizona. In: C. H. Lowe, ed. *The vertebrates of Arizona*, pp. 133–151. University of Arizona Press, Tucson.

Minckley, W. L. 1973. *Fishes of Arizona*. Sims Printing Co., Phoenix, Arizona.

Minckley, W. L., and D. E. Brown. 1994. Wetlands. In: D. E. Brown, ed. *Biotic communities of the American Southwest: United States and Mexico*, pp. 237–286. Utah State University Press, Logan.

Minckley, W. L., and J. E. Deacon. 1991. *Battle against extinction: Native fish management in the American West.* University of Arizona Press, Tucson.

Mitchell, D. L., D. Ivanova, R. Rabin, T. J. Brown, and K. Redmond. 2002. Gulf of California sea surface temperatures and the North American monsoon: Mechanistic implications from observations. *Journal of the Climate* 15:2261–2281.

Mitsch, W. J., and J. G. Gosselink. 1993. *Wetlands*. Van Nostrand Reinhold, New York.

Moir, W. 2006. *Vascular plants of the San Francisco Peaks.* Available via http://www4.nau.edu/deaver/SFP%20flora.htm

Moir, W. H., and J. A. Ludwig. 1979. *A classification of spruce-fir and mixed conifer habitat types of Arizona and New Mexico.* Research Note RM-37, U.S. Department of Agriculture Forest Service, Fort Collins, Colo.

Monson, G., and A. R. Phillips. 1964. Species of birds in Arizona. In: C. H. Lowe, ed. *The vertebrates of Arizona*, pp. 175–248. University of Arizona Press, Tucson.

Monson, G., and A. R. Phillips. 1981. *Annotated checklist of the birds of Arizona.* University of Arizona Press, Tucson.

Nations, J. D. 1989. Cretaceous history of northeastern and east-central Arizona. In: J. P. Jenney and S. J. Reynolds, eds. *Geologic evolution of Arizona*, pp. 435–446. Arizona Geological Society Digest 17, Arizona Geological Society, Phoenix.

Neary, D. G., K. C. Ryan, and L. F. DeBano. 2005. *Wildland fire in ecosystems: Effects of fire on soil and water*, Vol. 4. General Technical Report RMRS-GTR-42, U.S. Department of Agriculture Forest Service, Fort Collins, Colo.

Nelson, S. M., and D. C. Anderson. 1994. An assessment of riparian environmental quality by using butterflies and disturbance susceptibility scores. *Southwestern Naturalist* 39:137–142.

Nichol, A. A. 1937. *The natural vegetation of Arizona*, pp. 81–222. Technical Bulletin 68, Arizona Agricultural Experiment Station, University of Arizona, Tucson.

Nowak, R. M. 1974. *The gray wolf in North America: A preliminary report.* New York Zoological Society and the U.S. Bureau of Sport Fish and Wildlife, Washington, D.C.

O'Brien, R. A. 2002. *Arizona's forest resources, 1999.* Resource Bulletin RMRS-RB-2, U.S. Department of Agriculture Forest Service, Fort Collins, Colo.

Ort, M. H., T. A. Dallegge, J. A. Vazquez, and D. L. White. 1998. Volcanism and sedimentation in the Miocene-Pliocene Hopi Buttes and Hopi Lake. In: E. M. Duebendorfer, ed. *Geologic excursions in northern and central Arizona,* pp. 35–58. Field trip guidebook for the Geological Society of America Rocky Mountain Section Meeting, May 1998, Northern Arizona University, Flagstaff.

Patton, D. R., and I. B. Judd. 1970. The role of wet meadows as wildlife habitats in the Southwest. *Journal of Range Management* 23:272–275.

Pearson, G. A. 1920a. Factors controlling the distribution of forest types. Part I. *Ecology* 1:139–159.

Pearson, G. A. 1920b. Factors controlling the distribution of forest types. Part II. *Ecology* 1:289–308.

Pearson, G. A. 1933. The conifers of northern Arizona. *Plateau* 6:1–7.

Pearthree, P. A., and D. B. Bausch. 1999. Earthquake hazards map. Arizona Geological Survey Map 34, Scale 1:1,000,000.

Peirce, H. W. 1984a. The Mogollon escarpment. *Arizona Bureau of Geology and Mineral Technology, Fieldnotes* 14:8–11.

Peirce, H. W. 1984b. Some late Cenozoic basins and basin deposits of southern and western Arizona. In: T. L. Smiley, J. D. Nations, T. L. Péwé, T. L., and J. P. Schafer, eds. *Landscapes of Arizona: The geological story,* pp. 207–227. University Press of America, Lanham, Md.

Peizhen, Z., P. Molnar, and W. R. Downs. 2001. Increased sedimentation rates and grain sizes 2–4 Myr ago due to the influence of climate change on erosion rates. *Nature* 410:891–897.

Peterson, F. F. 1981. *Landforms of the Basin and Range province, defined for soil survey.* Technical Bulletin 28, Nevada Agricultural Experiment Station, University of Nevada, Reno.

Péwé, T. L. 1978. Terraces of the lower Salt River Valley in relation to the Late Cenozoic history of the Phoenix Basin, Arizona (scale 1:32,000). In: D. M. Burt and T. L. Péwé, eds. *Guidebook to the geology of central Arizona,* pp. 1–45. Special Paper No. 2, Arizona Bureau of Geology and Mineral Technology, Phoenix.

Philander, S. G. 1990. *El Niño, La Niña, and the Southern Oscillation.* Academic Press, San Diego, Calif.

Phillips, A. R., J. T. Marshall, and G. Monson. 1964. *The birds of Arizona.* University of Arizona Press, Tucson.

Phillips, S. J., and P. W. Comus, eds. 2000. *A natural history of the Sonoran Desert.* Arizona Desert Museum Press, Tucson.

Rasmusson, E. M., and J. M. Wallace. 1983. Meteorological aspects of the El Niño/Southern Oscillation. *Science* 222:1195–1202.

Reed, T. R. 1933. The North American high-level anticyclone. *Monthly Weather Review* 61:321–325.

Reed, T. R. 1939. Thermal aspects of the high-level anticyclone. *Monthly Weather Review* 67:201–204.

Reveal, J. L. 1979. Biogeography of the Intermountain region: A speculative appraisal. *Mentzelia* 4:1–92.

Reynolds, H. G., W. P. Clary, and P. F. Ffolliott. 1970. Gambel oak for southwestern wildlife. *Journal of Forestry* 68:545–547.

Rich, L. R., and J. R. Thompson. 1974. *Watershed management in Arizona's mixed conifer forests: The status of our knowledge.* Research Paper RM-130, U.S. Department of Agriculture Forest Service, Fort Collins, Colo.

Richard, S. M., S. J. Reynolds, J. E. Spencer, and P. A. Pearthree. 2000. Geologic map of Arizona. Arizona Geological Survey Map 35, Scale 1:1,000,000.

Rink, G. 2005. A checklist of the vascular flora of Canyon de Chelly National Monument, Apache County, Arizona. *Journal of the Torrey Botanical Society* 132:510–532.

Rinne, J. N. 1996. Short-term effects of wildfire on fishes and aquatic macrovertebrates: Southwestern United States. *Journal of Fisheries Management* 16:653–658.

Rinne, J. N. 2003. Flows, fishes, foreigners and fires: Relative impacts on southwestern native fishes. *Hydrology and Water Resources in Arizona and the Southwest* 33:79–83.

Rinne, J. N., and W. L. Minckley. 1991. *Native fishes of arid lands: A dwindling resource of the desert Southwest.* General Technical Report RM-206, U.S. Department of Agriculture Forest Service, Fort Collins, Colo.

Robbie, W. A. 2004. Grassland assessment categories and extent. In: D. M. Finch, ed. *Assessment of grassland ecosystem conditions in the southwestern United States,* Vol. 1, pp. 11–17. General Technical Report RMRS-GTR-135, U.S. Department of Agriculture Forest Service, Fort Collins, Colo.

Ropelewski, C. F., and M. S. Halpert. 1986. North American precipitation and temperature patterns associated with the El Niño/Southern Oscillation. *Monthly Weather Review* 114:2352–2362.

Scarborough, R. B. 1989. Cenozoic erosion and sedimentation in Arizona. In: J. P. Jenney and S. J. Reynolds, eds. *Geologic evolution of Arizona*, pp. 515–538. Arizona Geological Society Digest 17, Arizona Geological Society, Phoenix.

Schade, C. B. 2003. Factors associated with the distribution and abundance of non-native fishes in streams of the American West. Master's thesis, University of Arizona, Tucson.

Schade, C. B., and S. A. Bonar. 2005. Distribution and abundance on non-native fishes in streams of the American West. *Journal of Fisheries Management* 25:1386–1394.

Schubert, G. H. 1974. *Silviculture of southwestern ponderosa: The status of our knowledge.* Research Paper RM-123, U.S. Department of Agriculture Forest Service, Fort Collins, Colo.

Schwalen, H. C. 1942. *Rainfall and runoff in the upper Santa Cruz River drainage basin*, pp. 421–472. Technical Bulletin 45, Agricultural Experiment Station, University of Arizona, Tucson.

Schwarth, H. S. 1914. *A distributional list of the birds of Arizona.* Pacific Coast Avifauna Publication 10, Cooper Ornithological Club, Waco, Tex.

Sellers, W. D., ed. 1960. *Arizona climate.* University of Arizona Press, Tucson.

Sellers, W. D., and R. H. Hill. 1974. *Arizona climate, 1931–1972.* University of Arizona Press, Tucson.

Sellers, W. D., R. H. Hill, and M. Sanderson-Rae. 1987. *Arizona climate: The first hundred years.* Department of Atmospheric Science, University of Arizona, Tucson.

Shafiqullah, M., P. E. Damon, D. J. Lynch, S. J. Reynolds, W. A. Rehrig, and R. H. Raymond. 1980. K-Ar geochronology and geologic history of southwestern Arizona and adjacent areas. In: J. P. Jenney and C. Stone. *Studies in western Arizona*, pp. 201–260. Arizona Geological Society Digest 12, Arizona Geological Society, Phoenix.

Shantz, H. L. 1927. Drought resistance and soil moisture. *Ecology* 8:145–157.

Sheppard, P. R., A. C. Comrie, G. D. Packin, K. Angersbach, and M. K. Hughes. 2002. The climate of the U.S. Southwest. *Climate Research* 21:219–238.

Sheridan, M. F. 1984. Volcanic landforms. In: T. L. Smiley, J. D. Nations, T. L. Péwé, and J. P. Schafer, eds. *Landscapes of Arizona: The geological story*, pp. 79–110. University Press of America, Lanham, Md.

Shrader-Frechette, K. 2001. Non-indigenous species and ecological explanation. *Biology and Philosophy* 16:507–519.

Shreve, F. 1915. *The vegetation of a desert mountain range as conditioned by climatic factors.* Publication 217, Carnegie Institute, Washington, D.C.

Shreve, F. 1936. The transition from desert to chaparral. *Madrono* 3:257–264.

Singer, M. J., and D. N. Munns. 1996. *Soils: An introduction.* Prentice Hall, Upper Saddle River, N.J.

Smiley, T. L. 1984. Climatic change during landform development. In: T. L. Smiley, J. D. Nations, T. L. Péwé, and J. P. Schafer, eds. *Landscapes of Arizona: the geological story,* pp. 55–77. University Press of America, Lanham, Md.

Smith, H. V. 1956. *The climate of Arizona.* Bulletin 279, Agricultural Experiment Station, University of Arizona, Tucson.

Smith, P. B. 1970. New evidence for a Pliocene marine embayment along the lower Colorado River area, California and Arizona. *Geological Society of America Bulletin* 81:1411–1420.

Snyder, K. A., D. P. Guertin, R. L. Jemison, and P. F. Ffolliott. 2002. Riparian plant community patterns: A case study from southeastern Arizona. *Journal of the Arizona-Nevada Academy of Science* 34:106–111.

Soil Survey Staff. 1998. *Keys to soil taxonomy.* U.S. Department of Agriculture Natural Resources Conservation Service, Washington, D.C.

Spencer, J. E., and P. J. Patchett. 1997. Sr isotope evidence for a lacustrine origin for the upper Miocene to Pliocene Bouse Formation, lower Colorado River trough, and implications for timing of Colorado Plateau uplift. *Geological Society of America Bulletin* 109:767–778.

Spencer, J. E., and S. J. Reynolds. 1989. Middle Tertiary tectonics of Arizona and the Southwest. In: J. P. Jenney and S. J. Reynolds, eds. *Geologic evolution of Arizona,* pp. 539–574. Arizona Geological Society Digest 17, Arizona Geological Society, Phoenix.

Spencer, J. E., S. M. Richard, S. J. Reynolds, R. J. Miller, M. Shafiqullah, M. J. Grubensky, and W. G. Gilbert. 1995. Spatial and temporal relationships between mid-Tertiary magmatism and extension in southwestern Arizona. *Journal of Geophysical Research* 100:10321–10351.

Stewart, J. H. 1978. Basin-range structure in western North America: A review. In: R. B. Smith and G. P. Eaton, eds. *Cenozoic tectonics and regional geophysics of the western Cordillera,* pp. 1–31. Geological Society of America Memoir 152, Geological Society of America, Boulder, Colo.

Strahler, A. H., and A. N. Strahler. 1992. *Modern physical geography.* John Wiley & Sons, New York.

Stredl, M. J., and J. M. Howland. 1995. Conservation and management of Madrean populations of the Chiricahua leopard frog. In: L. F. DeBano, P. F. Ffolliott, A. Ortega-Rubio, G. J. Gottfried, R. H. Hamre, and C. B. Edminster, tech. coord. *Biodiversity and management of the Madrean archipelago: The sky islands of southwestern United States and north-western Mexico*, pp. 379–385. General Technical Report RM-GTR-264, U.S. Department of Agriculture Forest Service, Fort Collins, Colo.

Stromberg, J. C. 2002. Restoration of riparian vegetation in the south-western United States. *Journal of Arid Environments* 49:17–34.

Stromberg, J., M. Briggs, C. Courley, M. Scott, P. Shafroth, and L. Stevens. 2004. Human alterations of riparian ecosystems. In: M. B. Baker Jr., P. F. Ffolliott, L. F. DeBano, and D. G. Neary, eds. *Riparian areas of the south-western United States: Hydrology, ecology, and management*, pp. 99–126. Lewis Publishers, Boca Raton, Fla.

Stubbendieck, J., S. L. Hatch, and L. M. Landholt. 2003. *North American wildland plants: A field guide.* University of Nebraska Press, Lincoln.

Swetnam, T. W., and J. L. Betancourt. 1990. Fire–Southern Oscillation relations in the southwestern United States. *Science* 249:1017–1020.

Swetnam, T. W., and J. L. Betancourt. 1998. Mesoscale disturbance and ecological response to decadal climatic variability in the American Southwest. *Journal of the Climate* 11:3128–3147.

Sykes, G. 1931. Rainfall investigation in Arizona and Sonora by means of long-period rain gauges. *Geographical Review* 21:229–233.

Szaro, R. C. 1989. Riparian forest and scrubland community types of Arizona and New Mexico. *Desert Plants* 9:70–138.

Takhtajan, A. 1986. *Floristic regions of the world.* University of California Press, Berkeley.

Tecle, A., D. G. Neary, P. F. Ffolliott, and M. B. Baker Jr. 2003. Water quality in forested watershed of the southwestern United States. *Journal of the Arizona-Nevada Academy of Science (Special issue: Watershed management in Arizona)* 35:48–57.

Thill, R. E., P. F. Ffolliott, and D. R. Patton. 1983. *Deer and elk forage production in Arizona mixed conifer forests.* Research Paper RM-248, U.S. Department of Agriculture Forest Service, Fort Collins, Colo.

Thornthwaite, C. W. 1948. An approach toward a rational classification of climate. *Geographical Review* 38:55–94.

Thorud, D. B., and P. F. Ffolliott. 1973. *A comprehensive analysis of a major storm and associated flooding in Arizona.* Technical Bulletin 202, Arizona Agricultural Experiment Station, University of Arizona, Tucson.

Tidwell, W. D., S. R. Rushforth, and D. Simper. 1972. Evolution of floras in the Intermountain region. In: A. Cronquist, A. H. Holmgren, N. H. Holmgren, and J. L. Reveal, eds. *Intermountain flora: Vascular plants of the Intermountain West, USA*, Vol. 1, pp. 19–39. New York Botanical Garden, New York.

Titley, S. R. 1995. Geological summary and perspective of porphyry copper deposits in southwestern North America. In: F. W. Pierce and J. G. Bolm, eds. *Porphyry copper deposits of the American Cordillera*, pp. 6–20. Arizona Geological Society Digest 20, Arizona Geological Society, Phoenix.

Titley, S. R., and D. C. Marozas. 1995. Processes and products of supergene copper enrichment. In: F. W. Pierce and J. G. Bolm, eds. *Porphyry copper deposits of the American Cordillera*, pp. 156–168. Arizona Geological Society Digest 20, Arizona Geological Society, Phoenix.

Todd, T. N. 1976. Pliocene occurrence of the Recent atherinid fish *Colpichthys regis* in Arizona. *Journal of Paleontology* 50:462–466.

Troyo-Dieguez, E., F. de Lachica-Bonilla, and J. L. Fernandez-Zayas. 1990. A simple aridity equation for agricultural purposes in marginal zones. *Journal of Arid Environments* 19:353–362.

UNESCO (United Nations Educational, Scientific and Cultural Organization). 1979. Map of the world distribution of arid regions: Explanatory note and map. MAB Technical Notes 7, UNESCO, Paris.

U.S. Department of Agriculture Natural Resources Conservation Service. 2005. *The PLANTS database*. Available via http://www.plants.usda.gov. National Plant Data Center, Baton Rouge, La.

U.S. Fish and Wildlife Service. 1986. Endangered and threatened wildlife and plant species. *Federal Register* 51(126):237679–243781.

U.S. Fish and Wildlife Service. 2007. *Environmental Conservation Online System (ECOS)*. Available via http://ecos.fws.gov

Van Devender, T. R., and W. G. Spaulding. 1979. Development of vegetation and climate in the southwestern United States. *Science* 204:701–710.

Van Devender, T. R., R. S. Thompson, and J. L. Betancourt. 1987. Vegetation history of the deserts of southwestern North America: The nature and timing of the Late Wisconsin–Holocene transition. In: W. F. Ruddiman and H. E. Wright Jr., eds. *North America and adjacent oceans during the last deglaciation*, pp. 323–352. Geological Society of America, Boulder, Colo.

Wirt, L., and H. W. Hjalmarson. 2000. *Sources of springs supplying base flow to the Verde River headwaters, Yavapai County, Arizona*. Open-File Report 99-0378, U.S. Geological Survey, Phoenix, Ariz.

Wolfe, E. W. 1984. The volcanic landscape of the San Francisco volcanic field. In: T. L. Smiley, J. D. Nations, T. L. Péwé, and J. P. Schafer, eds. *Landscapes of Arizona: The geological story*, pp. 111–136. University Press of America, Lanham, Md.

Wolfe, J. A., C. E. Forest, and P. Molnar. 1998. Paleobotanical evidence of Eocene and Oligocene paleoaltitudes in midlatitude western North America. *Geological Society of America Bulletin* 110:664–678.

Wright, H. A. 1980. *The role and use of fire in the semidesert grass-shrub type.* General Technical Report INT-85, U.S. Department of Agriculture Forest Service, Ogden, Utah.

Wright, H. A. 1990. Role of fire in the management of southwestern ecosystems. In: J. S. Krammes, tech. coord. *Effects of fire management of southwestern natural resources*, pp. 1–5. General Technical Report RM-191, U.S. Department of Agriculture Forest Service, Fort Collins, Colo.

Wrucke, C. T. 1989. The middle Proterozoic Apache Group, Troy Quartzite, and associated diabase of Arizona. In: J. P. Jenney and S. J. Reynolds, eds. *Geologic evolution of Arizona*, pp. 239–258. Arizona Geological Society Digest 17, Arizona Geological Society, Phoenix.

Xu, J., X. Gao, J. Shuttleworth, S. Sorooshian, and E. Small. 2004. Model climatology of the North American monsoon onset period during 1980–2001. *Journal of the Climate* 17:3892–3906.

Young, R. A. 2001. The Laramide-Paleogene history of the western Grand Canyon region: Setting the stage. In: R. A. Young and E. E. Spamer, eds. *The Colorado River: Origin and evolution*, pp. 7–15. Grand Canyon Association Monograph 12, Grand Canyon, Ariz.

About the Contributors

D. Robert Altschul is an associate professor emeritus in the Department of Geography and Regional Development, University of Arizona. He received his PhD in geography from the University of Illinois. His teaching and research for 30 years focused on physical and human geography of the world's arid and semiarid lands. Recently, he has worked on a comprehensive overview of the physical, human, and economic characteristics of arid and semiarid regions for inclusion in the "Deserts of the World," a Website offering.

Owen K. Davis is a professor in the Department of Geosciences, University of Arizona. He is director of the Palynology Laboratory and a past-president of the Arizona-Nevada Academy of Science. His research emphasis is environmental changes in arid lands in the past 30 million years, particularly in the last 2000 years of human impacts on environments of the Southwest.

Leonard F. DeBano is a professor in the School of Natural Resources, University of Arizona. He previously was a project leader and soil scientist with the USDA Forest Service in California and Arizona, conducting research in watershed management. He has published widely on watershed management, fire effects on soils, and riparian management.

Peter F. Ffolliott is a professor in the School of Natural Resources, University of Arizona, with joint appointments in the Arid Lands Studies Program and the Laboratory for Tree Ring Research. He is a past-president of the Arizona-Nevada Academy of Science. His teaching and research interests focus on integrated watershed management including biophysical and socioeconomic factors relating people to their land and water. He has authored and coauthored over 500 publications on a diversity of topics relative to ecosystem-based multiple-use management of natural resources.

Gerald J. Gottfried is a research forester with the Rocky Mountain Research Station, U.S. Forest Service, Phoenix, Arizona. He is responsible for coordinating the field studies of the Southwestern Borderlands Ecosystem Management Project with the mission of implementing a comprehensive ecosystem management plan for the borderland region. His research emphasis is range restoration by mechanical methods and prescribed burning, erosion-sedimentation dynamics, and the ecological and managerial relationships of natural resources in the encinal (oak) savannas.

Paul R. Krausman was a professor of wildlife conservation and management in the School of Natural Resources, University of Arizona, from 1978 to 2007, when he joined the faculty at the University of Montana as the Boone and Crockett Professor of Wildlife Conservation. He has devoted his professional career to understanding the basic relationships between large terrestrial mammals and the alterations to their habitats caused by human actions. Most of his research has been in the southwestern deserts, but he also studied large ungulates in Egypt, India, and Mexico.

Steven P. McLaughlin is a professor emeritus of Arid Lands Studies and curator emeritus of the Herbarium at the University of Arizona. His primary research efforts have focused on comprehensive studies of quantitative floristic plant geography, economic botany, and plant community ecology.

Daniel G. Neary is a science team leader and research soil scientist, Southwest Watershed Team, Water and Air Program, Rocky Mountain Research Station, U.S. Forest Service, Flagstaff, Arizona. He holds degrees from Michigan State University and is currently an adjunct professor at the University of Arizona, Northern Arizona University, and the University of Florida. His current research is centered on the watershed-scale effects of fire on soils and water resources.

William D. Sellers is a professor emeritus of Atmospheric Sciences, University of Arizona. He was associated with the University of Arizona from 1957 through 1997, authoring or coauthoring several milestone publications dealing with Arizona's climate. Sellers is considered to be a pioneer in the studies of physical climatology and numerical climate modeling.

Jon E. Spencer received his PhD from MIT in 1981. For the past 25 years he has worked at the Arizona Geological Survey, where he is currently a senior geologist. His research has focused on the bedrock geology of the Basin and Range Province in southern and western Arizona. He has authored dozens of geologic maps and articles about Arizona's geologic features.

Figure Credits

2.1. The world's drylands and zones of climatic aridity. Adapted from Meigs 1953; UNESCO 1979; Dregne 1983.

3.1. A convective buildup of clouds. Photo by Cody L. Stropki.

4.1. A simplified geologic map of Arizona. Map by Jon E. Spencer.

4.2. Southwestern North America. Map by Jon E. Spencer.

4.3. The geologic time scale. Diagram by Jon E. Spencer.

4.4. A simplified geologic map of the Colorado Plateau region of northeastern Arizona. Map by Jon E. Spencer.

4.5. Schematic cross-sections showing the evolution of the Basin and Range Province and the Salton Trough. Drawing by Jon E. Spencer.

4.6. The Grand Canyon. Photo by Peter F. Ffolliott.

5.1. Beaver Creek watershed. Photo by Peter F. Ffolliott.

5.2. Sediment accumulation in a stream channel. Photo by Peter F. Ffolliott.

6.1. The paloverde-saguaro (*Parkinsonia-Carnegia*) association of the Sonoran Desert. Photo by Peter F. Ffolliott.

6.2. The creosote bush–burr ragweed (*Larrea-Ambrosia*) association of the Sonoran Desert. Photo by Peter F. Ffolliott.

6.3. Columnar cacti, including the saguaro. Photo by Cody L. Stropki.

6.4. The physiognomy of the Chihuahuan Desert. Photo by Peter F. Ffolliott.

6.5. The Joshua tree (*Yucca brevifolia*). Photo by Cody L. Stropki.

6.6. Landscape of the Great Basin Desert. Photo by Peter F. Ffolliott.

6.7. A variety of bunchgrasses and other species cover desert grasslands. Photo by Peter F. Ffolliott.

6.8. Plains grasslands intermingle with pinyon-juniper and sagebrush. Photo by Peter F. Ffolliott.

6.9. Trees and shrubs are not common in mountain grasslands. Photo by Peter F. Ffolliott.

6.10. Dense stands of sclerophyllous shrubs. Photo by Peter F. Ffolliott.

6.11. Emory oak (*Quercus emoryi*). Photo by Peter F. Ffolliott.

6.12. Pinyon-juniper woodlands of Arizona. Photo by Peter F. Ffolliott.

6.13. Scattered ponderosa pine forests. Photo by Peter F. Ffolliott.

6.14. Ponderosa pine forests of northern, central, and eastern Arizona are concentrated in a belt. Photo by Peter F. Ffolliott.

6.15. Engelmann and blue spruce intermingle with Rocky Mountain Douglas-fir and white fir. Photo by Peter F. Ffolliott.

6.16. Quaking aspen trees become established on disturbed sites. Photo by Peter F. Ffolliott.

6.17. The Engelmann spruce is the dominant tree species in the Arizona spruce-fir forests. Photo by Peter F. Ffolliott.

6.18. An isolated alpine plant community. Photo by Cody L. Stropki.

6.19. Low-elevation riparian associations. Photo by Peter F. Ffolliott.

6.20. Middle-elevation riparian associations. Photo by Peter F. Ffolliott.

6.21. Upper-elevation riparian associations. Photo by Alvin L. Medina.

6.22. Riverine wetland association. Photo by Cody L. Stropki.

7.1. The major floristic subdivisions of the state. Map by Steven P. McLaughlin.

7.2. Eight selected subprovincial floristic elements of North America. Maps by Steven P. McLaughlin.

7.3. Canyon de Chelly National Monument. Map by Steven P. McLaughlin.

7.4. The San Francisco Peaks area. Map by Steven P. McLaughlin.

7.5. The Greater Sedona area. Map by Steven P. McLaughlin.

7.6. The Cabeza Prieta National Wildlife Refuge. Map by Steven P. McLaughlin.

7.7. The Huachuca Mountains. Map by Steven P. McLaughlin.

7.8. Frequency distribution of ranges of native plant species in Arizona's flora. Graphs by Steven P. McLaughlin.

8.1. Streams, rivers, lakes, and reservoirs are habitats. Photo courtesy of the USDA Forest Service

8.2. Male mountain lion (*Puma concolor*). Photo by Kerry L. Nicholson.

8.3. Excavated nesting holes in dead or deteriorating trees. Photo by Peter F. Ffolliott.

9.1. Ponderosa pine (*Pinus ponderosa*) trees in stands. Photo by Peter F. Ffolliott.

Index